I dedicate this book to the memory of my parents,

RAY AND ZELDA MITCHELL

UNSIMPLE TRUTHS

UNSIMPLE TRUTHS

SCIENCE, COMPLEXITY, AND POLICY

SANDRA D. MITCHELL

THE UNIVERSITY OF CHICAGO PRESS · *CHICAGO AND LONDON*

The University of Chicago Press, Chicago 60637

The University of Chicago Press, Ltd., London

Paperback edition 2012

Printed in the United States of America

22 21 20 19 18 17 16 15 14 13 12   3 4 5 6 7

ISBN-13: 978-0-226-53262-2 (cloth)

ISBN-13: 978-0-226-00662-8 (paper)

ISBN-10: 0-226-53262-3 (cloth)

ISBN-10: 0-226-00662-X (paper)

Library of Congress Cataloging-in-Publication Data

Mitchell, Sandra D., 1951–

Unsimple truths : science, complexity, and policy / Sandra D. Mitchell.

p. cm.

Includes bibliographical references and index.

ISBN-13: 978-0-226-53262-2 (cloth : alk. paper)

ISBN-10: 0-226-53262-3 (cloth: alk. paper)

1. Complexity (Philosophy). 2. Science—Philosophy. I. Title.

Q175.32.C65M585 2009

501—dc22   2009015682

∞ This paper meets the requirements of ANSI/NISO Z39.48-1992 (Permanence of Paper).

CONTENTS

PREFACE

This book argues that our vision of the world, how it is constituted, what kind of knowledge we can have, how to investigate it, and how to act in light of the results of those investigations should reflect nature's complexity and contingency. Having been trained in the philosophy of science of Carnap, Popper, Lakatos, and Kuhn, I propose a new perspective, integrative pluralism, which has emerged, in part, from my witnessing both the strengths and the inadequacies of more rigid, unitary, and simple models of science in tracking new developments in the sciences of complex phenomena. My greatest teacher has been the work of contemporary biologists who grapple on a day-to-day basis with the complexities I describe in this work. I argue that new results, new techniques, and new challenges for science and society at large require a change in our understanding of science and nature. This book aims to encourage further reflection on foundational issues in philosophy of science in response to continuing developments in science.

The ideas and arguments in this book were published in a German translation in the edition unseld series under the title *Komplexitäten: Warum wir erst anfangen, die Welt zu verstehen* (Suhrkamp Verlag, 2008). I have received very useful feedback about the text from German readers, colleagues, students, and friends, which has spurred me to make some changes in the style of presentation and content. In particular Peter K. Machamer, Jim Bogen, Ulrich Krohs, Thomas Cunningham, Yoichi

Ishida, and Jason Byron provided helpful comments on the first version. Thanks also to Julia Bursten for compiling the index. Joel M. Smith read and reread the developing text and set a high standard for clarity and accessibility. In addition, responding to suggestions from the three anonymous reviewers and my editor, Karen Merikangas Darling, at the University of Chicago Press, improved the clarity and structure of the text. My work has developed in a community of scholars whose insights have inspired me, especially the work of Nancy Cartwright, John Beatty, Stuart Kauffman, Elliott Sober, William Wimsatt, James F. Woodward, and Robert E. Page Jr.

I wish to thank the School of Arts and Sciences at the University of Pittsburgh for supporting an environment that is both stimulating and challenging, and the American Philosophical Society for the award of a sabbatical fellowship that permitted some undistracted time with which to concentrate my efforts on this project.

This book would not have been written at all without the continuing and unfailing support of my husband, intellectual critic, and friend, Joel M. Smith.

1

INTRODUCTION

Complexity is everywhere. From unicellular behavior to processes on a global or cosmological scale, nature presents us with astonishingly complex, interconnected patterns of behavior. The challenge of natural complexity, in its many forms, has been clear since the writings of the earliest scientists. The history of science has in large part been shaped by efforts to explain and predict in the face of this reality. Much of science, especially prior to the latter half of the twentieth century, adopted strategies involving reductive explanations designed to simplify the many complexities of nature in order to understand them. The earliest reductive approaches were ontological and, to the modern mind, naïve. They involved, for example, suggestions by the early "monists" that everything in nature was really made of one substance (for pre-Socratic philosophers, water was one candidate, air another; Kirk, Raven, and Schofield 1983). Later reductivist approaches were much more sophisticated and proved quite successful, for example, Isaac Newton's hypothesis that all complex motions could ultimately be reduced to three simple laws. Often complexity can in fact be reduced to a surprisingly simple explanation. One need only consider the reduction of the complex motions of the planets as viewed from Earth to the comparatively simple Copernican and Keplerian explanations to find validation for the drive to explain the complex in terms of the simple. Philosophers of science have, correspondingly, offered accounts of how reductivist strategies constituted the best ways to understand and predict nature: how they are the very model of rationality.

However, recent developments in the scientific study of new types of complex properties and behaviors do not fit comfortably into the structures of explanation, experimental inference, and prescriptions for rational policy formulation that have been developed by centuries of philosophical analysis. For example, irreducible but explicable complexity practically defines global climate change and the ways in which human actions may influence it.[1] The development of a multicellular organism from a single cell and the dynamics of a honeybee colony prove surprisingly resistant to the traditional strategies. Why? One reason is that the philosophical focus for several centuries has been primarily directed to the scientific achievements of physics. Physics is a domain in which a reduction of complex phenomena to simpler ones has been particularly successful. However, many complex behaviors in biological and social sciences seem not to yield as well to a reductive approach. Are these sciences just immature? Are they not sciences at all because their explanations are of a different nature than those in physics? Or are the philosophical accounts of scientific rationality perhaps too limited to account for recent developments in modern science? As I will discuss throughout this work, there is a mismatch between the current sciences of complex behaviors and standard philosophical criteria for scientific knowledge. It is my view that new approaches to knowledge and action are required to understand and manage the complexity of nature. These practices can be found in contemporary sciences. My philosophical view, which I call integrative pluralism, explains how these modern scientific practices can be well grounded in an expanded epistemology of science that embraces both traditional reductive and new, multilevel, context-dependent approaches to scientific explanation and prediction.

One quite successful reductivist strategy in science has been the search for fundamental laws. The search for universal, exceptionless laws, for example, was taken by nineteenth-century British philosophers (Herschel 1830; Whewell 1840; Mill 1843) to be *the* goal of scientific investigation, as they reflected on the enormous success of Isaac Newton's laws of motion and universal gravitation. Yet if all of today's scientists restricted themselves to the search for and use of generalizations that meet the stringent standards of universality and exceptionlessness, much of great value that has been discovered about complex, contingent, and evolved structures would not qualify as scientific knowledge.

Similarly, traditional philosophical analyses of the cleanest, clearest cases of causal interactions, perhaps the first step in philosophical analyses of science, are difficult if not impossible to apply to the messy, murky causal relations that are displayed by genes and phenotypes, human interventions on the global climate, or multilevel, feedback-laden phenomena studied in modern psychiatry. A revised and expanded epistemology is required to face the challenge of understanding these kinds of complex behaviors. This book builds on the critiques and proposals of philosophers who have challenged the hegemony of an overly limited philosophical conception of scientific practice. To begin to meet the challenge of an expanded vision, new questions must be addressed. How are complex structures different from simple ones? What types of complexity are there? How do the answers to these questions bear on our understanding of nature, how we come to know it, and how we should base our actions and policies on that knowledge? In what follows I will present arguments that the contingency of complex structures impels us to modify our conception of the character of usable knowledge claims beyond the narrow domain of universal, exceptionless laws. Exceptions are the rule and limitations are to be expected in the laws that describe many complex behaviors. We will see that the traditional notions of controlled experimentation as the best scientific method for ascertaining causal structures can fail to accommodate robust, dynamic behaviors. Both the search for "the" scientific method and for some small set of fundamental laws that explain everything, everywhere, must be replaced by a more pragmatic and pluralistic approach to scientific practice.

Perhaps the most important location in which a better understanding of complexity calls for a revision of our thinking is in making decisions and crafting policies that help navigate the complex structures that populate the world in which we live. Complexity often carries with it a type of ineliminable or "deep" uncertainty that is not adequately represented by methods better suited to more certain, predictable, and static parts of nature. Some of the uncertainty rests in ignorance of the many factors that contribute to complex processes, like the global climate, yet some is due to the role of chance or chaos affecting the process itself. Once-and-for-all policies based on consensual measures of quantitative risk just do not fit the bill. To try to force decisions about complex domains of deep uncertainty into inadequate models is wrong for many reasons. Most importantly, it can lead to ignoring the scientific knowledge we do have and prevent our policies from being informed by that science. I will discuss new means of modeling uncertainty and designs for responsive policies that are better at permitting us to use the best information we have to make the best decisions we can.

This book is a jointly normative and descriptive project. Adequacy of philosophical accounts of explanation or causal inference or rational choice are, for my strategy, constrained, but not determined, by the practices and insights of the best of contemporary science. What kinds of properties and processes are explanatory is in part a reflection of the discoveries and theories science continues to provide. This is an old story for the naturalist philosopher of science, whose raison d'être is to come to grips with epistemological problems by means of analyzing the best source of knowledge of nature, namely, science itself. In turn, science's reflection on its own practices should engage conceptual tools provided by philosophical analysis. In short, science speaks to philosophical concerns by its very practice, and philosophy speaks to scientific concerns by articulating implicit assumptions and providing conceptual clarification.

What follows is an excursion into the world of the complex, with forays into the problems that science confronts, and detours into the land of philosophical reasoning. This is a dialectic ride that brings the many kinds of complexity—compositional, dynamic, and evolved—into the forefront for both scientific and philosophical scrutiny. The result, I argue, is a warrant for expanding our epistemological view to embrace pluralism, pragmatism, and the deeply dynamic character of both the world and our knowledge of it. My argument is *not* one for a wholesale replacement of the traditional views but, rather, for an expansion of traditional epistemologies of science to accommodate aspects of knowledge that do not fit the older formulations.

I start with an example of the kind of complexity that motivates my study, some recent scientific research on the relationship between genes and major depressive disorder. I then turn to three areas for exploration: concepts, including those of emergence and the contingent character of scientific laws; investigations, engaging the problems for causal inference and our understanding of cause itself raised by dynamically reorganizing genetic networks; and finally to actions, where policies about genetically modified foods and global climate change require a rethinking of the best way to represent and act in light of the deep uncertainty that accompanies these forms of complexity.

To begin to understand many aspects our complex world, I will argue that we need to expand our conceptual frameworks to accommodate contingency, dynamic robustness, and deep uncertainty. The truths that attach to our world are not simple, global, and necessary, but rather plural and pragmatic. Holding scientific practice to the wrong standards diminishes the value that science can and does provide to our understanding of nature and to our success in acting in ways to further our individual and collective goals.

A Case of Complexity

Major depressive disorder is one of the most common psychiatric maladies afflicting adults and children, and is widely distributed among various ethnic and socioeconomic groups. In a recent European study, major depression was identified as one of the two most common psychiatric disorders, with almost 13% of people surveyed reporting a lifetime history of depression, and around 4% experiencing a depressive episode in the past twelve months (Alonso et al. 2004). In the United States, the chance of a person suffering from clinical depression in their lifetime is around 16%. In any given year, about 7% of the U.S. population experiences the debilitating symptoms of this disease (Kessler et al. 2003).

Clinical depression is not the expected sadness that is an appropriate response to tragic life events, like the death of a loved one, but rather a complex set of biochemical, chemical, neurological, neuroanatomical, psychological, and physical states. It is diagnosed by an inventory of symptoms including having a depressed mood or loss of interest and pleasure in almost all activities for a period of at least two weeks. These are often accompanied by sleep disturbances, appetite disturbances, changes in energy levels, difficulties with thinking and concentration, and sexual difficulties. Depression can be mild or severe, but it is a serious disorder the occurrence of which interferes with a person's usual behavior and functioning. The psychological toll is great, as the subjective experience or phenomenology of depression includes feelings of hopelessness, worthlessness, and guilt.[2]

Why do some people suffer from major depressive disorder and others do not? Why does a person experience the debilitating symptoms of the disease at a given time, in a given year, and not at some other time? How do we explain the causes of depression so that we can develop more effective treatments? These are the questions facing scientific researchers and clinicians trying to develop the best practical policies for the treatment of depression.

The DSM-IV (*Diagnostic and Statistical Manual of Mental Disorders*, fourth edition) identifies various factors, no one of which is necessary, that may be implicated in generating a major depressive episode. There is no diagnostic laboratory test that definitively shows if a person has or does not have depression. Rather, the disorder is characterized as one that may involve a "dysregulation of a number of neurotransmitter systems . . . alternations of several neuropeptides . . . [and for some] hormonal disturbances" (DSM-IV, 353). Functional magnetic brain imaging documents changes in cerebral blood flow and metabolism for some depressed patients, and sometimes changes in brain structure. Yet, according to the DSM-IV (353), "none of these changes are present in all individuals in a major depressive episode . . . nor is any particular disturbance specific to depression." In addition, there is a familial pattern with major depression. It is 1.5 to 3 times more common among first-degree biological relatives of individuals with this disorder compared with the general population (DSM-IV, 373).

In 2006 Kendler, Gardner, and Prescott presented a comprehensive top-down analysis of the etiology of major depressive disorder. Their 2006 study targeted depression in men, complementing a similar 2002 study of women. The combined research studied nearly 2,000 female twins and 3,000 male twins. The general conclusion was that depression is an etiologically complex disorder that involves multiple factors from multiple domains acting over developmental time (Kendler, Gardner, and Prescott 2002, 2006). Three distinct pathways for development were identified for adult depression, which involved genetic risk factors, anxiety, and conduct disorder as well as parental behavior, childhood abuse, and socioeconomic conditions. They concluded that "major depression is a paradigmatic multifactorial disorder, where risk of illness is influenced by a range of factors" (2006, 115).

This is the frustrating reality: at this point, it appears naïve to believe there is a "gene for depression," which would explain the malady through a traditional reductivist strategy. Major depressive disorder is a complex behavior of a complex system that is dependent on multiple causes at multiple levels of organization (chemical, physical, biological, neurological, psychological, and social). In recent years much has been learned about the components that are associated with this disorder at all these levels (see Schatzberg 2002). An important question that remains to be answered is, what are the relationships between the properties and behaviors at the different levels? It is worth going into some of the details of recent studies, as this case of complexity illuminates the motivation for the major thesis of this work, that new approaches to knowledge and action are required to understand and manage complex systems. Current scientific developments in the explanation and treatment of major depressive disorder provide an example of why a richer epistemology and new strategies for action are necessary.

That there are genetic risk factors for depression is indicated by the familial relationships and twin studies that show a higher probability of depressive disorder among related adults. Behavioral genetic studies are designed to analyze the variation in a condition, like major depressive disorder, in a population. Because twins can be either genetically identical (monozygotic) or fraternal (dizygotic), comparisons among these groups can help estimate how much of the difference in the incidence of the behavior is due to genetic factors and how much is due to environmental factors. The observational studies do indicate a genetic factor for major depressive disorder.

The main class of drugs used to treat depression is selective serotonin reuptake inhibitors (SSRIs), which target the midbrain serotonergic neurons. The activity of these neurons is controlled, at least in part, by what are known as 5-HT autoreceptors (Haddjeri and de Montigny

1998). A gene involved in the development of these receptors, called the 5-*HTT* gene, was identified in the 1990s (Bengel et al. 1997). The organization of the gene includes a promoter, or regulatory region, which was found to occur in two forms in human populations, one with a long promoter region and the other with a short promoter region. Since then studies have targeted further understanding the role of the 5-*HTT* gene (or serotonin transporter gene) in contributing to various psychiatric disorders. Hariri et al. (2005) investigated the relationship between the short allele of the 5-*HTT* gene and response to situations of environmental threat in the human amygdala, the part of the brain important for emotional and social behavior including normal fear and pathological anxiety. The results of their study, conducted on healthy subjects, indicated a correlation between the short allele and amygdala response to threat, but did not lead, as one who is trying to find the *physical cause* of depression might hope, to predictions of depressive mood. They concluded that the 5-*HTT* gene may be best classified as a susceptibility factor in depression, but not a sole determinant.

By looking at the biochemical, neurological, and genetic levels of organization, it is increasingly clear to the scientists investigating these phenomena that depression is not the result of a simple unilateral cause, nor is it a behavior of a system that is simply organized. That is, it appears that not only is there no single cause of major depressive disorder, but there is also no unique set of causes that, in aggregate, produce the effect. Thus what is commonly known in the philosophical literature as a "greedy reductionist strategy" (Dennett 1995), where information about the component genes, and nothing else, would lead to predictions of the likelihood of a depressive episode, is not going to be successful. There is no "gene for" depression (Kendler 2005a).

To add to the potential complexity of this phenomenon, recent studies indicate that the plasticity and variability of depression show interactions between the basic components, the genes, and the wider environmental context. Caspi et al. 2003 (see Zammit and Owen 2006 for review and critique) reported results of a prospective longitudinal study of 1,037 children regularly assessed from three years of age to twenty-six. They were divided into three groups based on their 5-*HTT* genotype: those with two copies of the short allele, those with one copy of the short allele and one long allele, and those with two copies of the long allele. They also measured incidence of stressful life events, for example, death of a parent or abuse, for the three groups. Their results showed that stressful life events would statistically (but not with certainty) predict depression in the groups who had at least one short allele but not for the group with two long alleles. They concluded that "*The 5-HTT gene interacts with life events* to predict depression symptoms, an increase in symptoms, depression diagnoses, new-onset diagnoses, suicidality, and an informant's report of depressed behavior" (Caspi et al. 2003, 387; my emphasis). Having the short *5-HTT* gene is not sufficient to predict depression, and having stressful life events is not sufficient to predict depression, but the two together interact such that having both the gene and the stressful environment do predict a higher probability of getting the disease in adults.[3]

Kendler et al. (2005) replicated the study, confirming the interaction of genes and environment.[4] They explored a mechanism by which having the gene might increase the probability of depression by making the individual more sensitive to stressful life events than those without the gene. They found that individuals with one or two short *5-HTT* alleles experienced a wider range of environmental events as stressful than did individuals with two long *5-HTT* alleles. What is significant about the gene-environment interaction studies is that they further support a conclusion that a nonreductionist approach (in the sense of not focusing exclusively on the most basic physical components of a system) to explaining the complex causal network leading to depression is necessary. They also suggest the possible presence of feedback loops between physical propensities and life events.

In general, psychiatric disorders are unlikely to be amenable to purely or even partially reductionist strategies. Because evidence suggests that they are behaviors where basic components interact with components at higher levels or with the external environmental context, an integrative methodology is needed to understand their causal history.[5] This is not to deny the role that genes play in the complex causal network, but rather to understand that role as context sensitive and the system in which it operates as displaying a high degree of contingent plasticity. As Eaves et al. (2005, 62) said, "The classical models of inherited (or 'Mendelian') disorders, which are caused by alleles at one or two loci with very little environmental influence, do not apply to most complex psychiatric disorders. The number of genes may be large, their effects may be small, and their effects on the phenotype may be many and varied as a function of other contingencies, including those of chance and the environment."

The weakness of the association between genes and psychiatric disorders speaks more to the variability of causal pathways and multiplicity of contingent factors than to the strength of genes in eliciting the disease phenotype.[6] Under a specific range of internal and external factors, it may well be that having two short alleles at the 5-*HTT* locus is what would make the difference between suffering from depression or not. But if that range of internal and external factors normally varies widely in a population, then knowing just the gene information will be insufficient to predict or explain the occurrence of depression. In context-rich causal scenarios, factors that have traditionally been allocated the role of cause, like the gene, and the factors that traditionally have not been a focus of research, but rather relegated to the general amorphous context or background conditions, will conspire to produce the outcome of real interest.[7] Most psychiatric disorders share the multilevel, multicomponent profile.

What we see in major depressive disorder is actually characteristic of many complex behaviors in nature in general. The behavior is associated with multiple levels of organization, from gene, to cell, to region of the brain, to hormonal systems, to affect and behavior (Craver 2007). In the case of depression, these features of complex systems have important implications for how one can study the etiology of the disorder as well as the kinds of knowledge one should expect to glean from such studies. As we will see, analogous issues arise in other domains of complex phenomena where causal explanations reflect similar context sensitivity and lack of uniqueness that we see in the case of depression.

Indeed, depression, like other psychiatric disorders, presents us with a quintessential example of some kinds of complexity. The challenge of understanding "how . . . many different small explanations fit together" (Kendler 2005b, 435) is a challenge I believe confronts many different sciences in many different ways. It is a fundamentally different challenge than the one framed for science by seventeenth- and eighteenth-century scientists and philosophers to find the few large explanatory principles that will explain all of nature. The world is just too complex to fully accomplish that goal.[8]

Shifting Paradigms in Epistemology

When people hear the word "complexity," they respond in different ways. Some think "complicated" or "messy," not being able to see the forest for the trees. Others think of a clutter of matter going this way and that with no chance to get a purchase on its behavior, to take hold of the "blooming, buzzing confusion" (James 1890, 462). Others think "chaos," in the traditional sense, something unrestrained and uncontrollable, a realm of unpredictability and uncertainty that doesn't yield to human understanding. None of these interpretations does justice to the tractable, understandable, evolved, and dynamic complexity that contemporary sciences say aptly characterizes our world.[9] Neither its complications nor its chaotic dynamics should scare away the curious, nor drive them to replace a clear-eyed investigation of the nuanced beauty of complexity with the austere, clean lines of the simple and timeless.

The world is indeed complex; so, too, should be our representations and analyses of it. Yet science has traditionally sought to reduce the "blooming, buzzing confusion" to simple, universal, and timeless underlying laws to explain what there is and how it behaves. The successes of the Scientific Revolution of the seventeenth century in providing simplifying, unifying representations, in particular Newton's laws of motion and his law of universal gravitation, led philosophers to define what they would admit as reliable knowledge in like terms.[10] In particular, nineteenth-century philosophers of science John Herschel (1830), William Whewell (1840), and John Stuart Mill (1843) attempted to capture the essence of "the scientific method" they attributed to Newton as a nearly algorithmic procedure for revealing the simplicity underlying the complexity of our daily experience. Although these three philosophers differed in many respects, they shared the view that the *vera causa*, or true causes of the phenomena we observe, would be described by simple underlying laws (see Snyder 2006). As Whewell put it, "The Consiliences of our Inductions give rise to a constant Convergence of our Theory towards Simplicity and Unity" (Butts 1989, 159). And John Stuart Mill said the "most perfect mode of investigation" is one that "tries its experiments not on the complex facts, but on the simple ones of which they are compounded" (1843, book 6, chapter 5). Universality, determinism, simplicity, and unification were designated the hallmarks of reliable knowledge, built from the solid foundation of empirical fact by means of induction.

But the world of Newtonian science did not persist. Twentieth-century physics challenged some of its most fundamental assumptions (see Cushing and McMullin 1989). More recently, as I describe in this book, the historical and contingent complexities of biological nature challenge the lingering hegemony of traditional views of scientific knowledge. Many have written on the impact of the discoveries of twentieth-century physics (e.g., Kragh 1999; Laughlin 2005; Smolin 1997, 2006). Fewer have explored new insights that should now emerge from recent work on biological complexity (but see Rosen 1985; Horan 1989; Emmeche 1997; Giere 1999; Wimsatt 2007; Bechtel and Richardson 1993; Kauffman 1984, 1993, 1995, 2008; Strevens 2003). This book is an investigation of how to extend and revise our epistemological framework in light of recent developments in the science of complex biological systems. I believe the lessons are applicable in all sciences of the complex.

Challenges for traditional epistemology are found in investigation of the cutting-edge science of complex systems. The current state of science tells us how and why the traditional epistemological framework is incomplete. However, it is not the case that the traditional framework always fails; there remain stunning successes since the time of Newton to testify to its power. The problem is that much of the world escapes its concepts and methods. This book aims to make explicit where it fails as well as to articulate the features of a new approach that will extend the scope of what counts as reliable knowledge of our complex world. I will suggest how an augmented epistemology, one that I have called "integrative pluralism," can accommodate our understanding of both simple and complex systems. I will also spell out the positive account of some of the details of the new epistemology in particular areas of focus in the sciences of the complex. Exploring more implications of taking an integrative pluralist approach constitutes a larger research program for philosophy of science.

My thesis is that complexity—for example, biological complexity—is not beyond our understanding; it requires new ways of understanding. It requires, in many cases, a more explicit and detailed analysis of the many roles context plays in shaping natural phenomena. It means that conditions often relegated to the status of "accidents" or "boundary conditions" be elevated to the subject of scientific study. Historical contingency conspires with episodes of randomness to create the actual forms and behaviors that populate life on our planet. Life is not simple, and our representations of life, our explanations of life, our theories of how life works, will not be simple either.[11] Features of the expanded approach to epistemology incorporate

- *Pluralism, integrating* multiple explanations and models at many levels of analysis instead of always expecting a single, bottom-level reductive explanation.
- *Pragmatism* in place of absolutism, which recognizes there are many ways to accurately, if partially, represent the nature of nature, including various degrees of generality and levels of abstraction. Which representation "works" best is dependent, in part, on our interests and abilities.
- The essentially *dynamic and evolving character of knowledge* in place of static universalism. This feature requires us to find new means of investigating nature and a reconfiguration of policy strategies for acting on the knowledge we obtain.

I will argue for a pluralist-realist approach to ontology, which suggests not that there are multiple worlds, but that there are multiple *correct* ways to parse our world, individuating a variety of objects and processes that reflect both causal structures and our interests. The view that there is only one true representation of the world exactly mapping onto its natural kinds is hubris. Any representation is at best partial, idealized, and abstract (Wimsatt 1987; Cartwright 1989). These are features that make representations usable, yet they are also features that limit our claims about the completeness of any single representation.

In the past I have described a continuum of abstraction (Mitchell 2000) that recognizes a type of ontological pluralism (see also Cartwright 2000; Longino 2002; Dupré 1993; Galison and Stump 1996; Kellert, Longino, and Waters 2006). This position supports the view that there are multiple correct and useful ways to describe the world, while completely rejecting any form of naïve relativism: not every posited description will be either correct or useful. Standards that justify a given representation consist of a combination of measures of predictive use, consistency, robustness, and relevance. These very same measures should then be used to guide our application of representations for understanding, prediction, and action.

Pluralism is the appropriate stance toward the models, theories, and explanations proffered by scientists. No longer should an a priori commitment to explanatory reduction to a set of basic entities that contemporary physics deploys, often successfully in a particular domain of inquiry, justify a devaluation of explanations appealing to properties at higher levels of organization. Some of the higher-level properties *emerge* in systems displaying complex interactions in the world. One undertakes such a devaluation of different forms of explanation at the peril of ignoring developments in contemporary science where the idea of "emergent properties" has recently resurfaced as a significant explanatory category. Indeed, I will articulate and defend a modern notion of emergence of natural phenomena that is explanatory and causally legitimate, but at the same time is not inconsistent with some form of compositional bottom-up explanation. The rejection of explanation by reduction to a lowest level of organization carries with it a new problem, namely, how are the different levels—for example, physical, chemical, genetic, neurological, and contextual—related or interacting in producing the behavior of a complex system.

Another aspect of pluralism that is part of the expanded epistemology is related to scientific laws. The nineteenth-century framework was designed in reflection on Newton's successes that a natural law must be a naturally necessary, universal, exceptionless truth. However, analysis of the types of generalizations that are common in sciences other than fundamental physics (and possibly even there; see Earman, Roberts, and Smith 2002; Cartwright 1994), shows that most accepted generalizations about the world are contingent, of limited scope, and exception-ridden. In the old framework there were only two classifications for true scientific generalizations: laws or accidental truths. Rather than lump most of what is known in chemistry, biology, psychology, and the social sciences into the class of accidental truths (to share space with claims of such limited use as "All the coins in Goodman's pocket are silver," the classic example of a truth expressed with the syntax of a universal generalization that would not be classified as a law, from Goodman 1947), what is needed is an expansion of the conceptual space of laws to reflect the different degrees of contingency, stability, and scope of the causal structures science represents. In the expanded epistemology of integrative pluralism, there are more kinds of things than fundamental physical particles, and there are more kinds of laws than naturally necessary, universal, exceptionless ones.

In integrative pluralism, pragmatism replaces absolutism in filling out the standards by which any one of the many truths about the world enter into scientific knowledge. The rejection of unity via reduction for all phenomena opens our view to accept that multiple representations can accurately describe different aspects of a given causal structure. What will direct the acceptance of any particular representation as appropriate for a particular inquiry or action are, in part, pragmatic interests: the goals we are trying to achieve by using the representation. Without the demand for reduction to one, most basic, material description, which level of abstraction to adopt is not determined in advance. For example, some patterns are visible only when material details are ignored. Darwin's insight into the causal structure responsible for adaptedness depended on recognizing similarity among extremely different systems, like the varieties of cowslip and Galápagos finches. Nearly all material properties of the two populations are different. Darwin's theory required abstracting away from matter to structure, seeing that the systems shared a relationship between the relative fitness of the surviving members of the populations and their respective environments. In order to explain species adaptedness, Darwin proposed that natural selection operating on heritable variations would cause individuals with any slight advantage to persist, whether the advantage is taller stalk, thicker beak, or darker pigmentation (see Mitchell 1993).

Additional pragmatic features, like cognitive accessibility or interest in elimination of undesirable effects in contrast to the production of desirable effects, will also contribute to the appropriateness of different scientific representations. Consider, for example, what kinds of factors need to be represented and at what level of abstraction if one wants to develop treatments for a disease. MacMahon, Pugh, and Ipsen (1960, 2) urge epidemiologists to seek out the "necessary," though rarely sufficient, causal conditions most easily managed by practical interventions rather than engage in "semantic exercises aimed at hierarchic classification of causes" (see also Krieger 1994). Clearly if one eliminates a necessary component in what might be a complex network of causes, the normal effect will not ensue. To get rid of malaria in a population, for example, it may be enough to drain the swamps and remove the mosquito that serves as the vector for carrying the disease into humans. If that is the purpose, there is no need to provide a detailed account of the contributions of the genetics of sickle cell to acquiring malaria (see also Gannett 1999). On the other hand, if one's purpose is to produce an effect that is the result of a complex set of individually insufficient but necessary parts of a causal complex that may be itself unnecessary but locally sufficient to produce the effect (see Mackie 1965), then one requires all the contributions to be represented and realized. If it is the case that failing to correctly fix the value of each of the many components of a complex set of requirements will result in failure to generate the effect, clearly the level of abstraction required will be more fine grained. A crude scale that captures a necessary component, but perhaps does not distinguish among the other interacting components whose inputs are irrelevant when the necessary component is removed, will not work when trying to produce a specific effect.

Finally, in integrative pluralism, recognition of the dynamic and evolving character of knowledge replaces an expectation of static universalism. As the universe has evolved, and as life on our planet has evolved, new casual structures have arisen. New elements were created in the evolution of stars; new forms of individuality, reproduction, and strategies for survival resulted from the processes of speciation. Our knowledge of the causal structures defined by the new systems evolves with the changing world. Not all causal structures are equally historically contingent: some were fixed in the first three minutes after the big bang (Weinberg 1993), and others are more recent and ephemeral, like retroviruses or human social arrangements. Some complex arrangements introduce new interactions that give rise to new structures that obey new laws. For example, there is good reason to believe that division of labor, a key component in sociality, initially arose through mechanisms of self-organization of interacting solitary insects, and only later were its features fine-tuned by natural selection at the colony level (Page and Mitchell 1998). Our world is changing, and our knowledge of it must change in response. A prima facie demand for a static, universal, exceptionless body of knowledge is, quite simply, a mismatch. With the appropriate dynamic picture of complexity come additional consequences for epistemology.

How do we investigate a world of entities behaving in a manner halfway between Parmenides' image of a static universe and Heraclitus's image of an ever-changing flow of properties?[12] Our frameworks for knowledge and action based on that knowledge both must change. How do we decide what policies to adopt based on the best science of our day in order to promote our individual and collective goals? Context-rich causal networks, chaotic determinism, and ineliminable uncertainty require new strategies for experimentation and decision making. Certainty or even estimable probability of predicted outcomes gives way to representations in terms of multiple-scenario projections. Once-and-for-all policy decisions need to be replaced by adaptive management schemes, which require monitoring, updating, and revision of actions on an ongoing basis (Murray and Marmorek 2003).

What makes the time ripe for an expanded approach to epistemology is not just the problems for the old framework. Expansion of explanatory and predictive methodologies is enabled by changes in the cognitive and representational tools at science's disposal. The scientific community now uses computational techniques that permit new representations of complexity. No longer are we restricted to what we can imagine and figure in our heads, or to mathematical representations that admit of closed-form solutions. Techniques for purely computational solutions and attendant visualization of complex differential equations and other mathematical representations of the world are now commonplace. With advanced computing comes the ability for generating simulations of systems with varying parameter values and varying contextual influences. No longer do we have to wait to construct a physical experiment to embody those values. Much more data and many more interactions can be accommodated. Indeed, computational tools permit us to store large amounts of information required for applying nonuniversal, exception-limited laws to new contexts. Instead of eternally true universal laws applicable to all space and time, an epistemology tuned to complexity will produce a host of contingent, domain-restricted generalizations that describe more or less stable causal structures and whose application requires the deployment of more data than a single human brain (or even large scientific communities) can manage. The rise of computational tools and representations makes an epistemology of integrative pluralism cognitively tractable.

This book is designed to begin the discussion of an expansion and revision of the traditional views of science and knowledge, codified in the nineteenth century by English philosophers, trying to make all scientists into new-age Isaac Newtons. These perspectives have dominated philosophical reflections on science. In what follows, I target three areas of human thought and practice where complexity requires us to revise old conceptions of how to reason and act rationally. I will explore how the complexity and contingency of natural processes changes

- how we *conceptualize* the world,
- how we *investigate* the world, and
- how we *act* in the world.

The examination of scientific cases in this book that present problems for the old epistemological framework will define conditions that must be met by integrative pluralism. It should be able to provide an expanded epistemological framework in which these problems are addressed. Rather than continuing to force the current state of scientific knowledge into what has become a straitjacket of a reductive, fundamental, monistic view of science, the very image of what should count as legitimate science itself must change as our conceptual framework expands. The picture of necessary universal laws that predict what will happen anywhere, anytime, and correlatively explain what has happened anywhere, anytime, is not an unreasonable methodology, but it is incomplete. It must be replaced by expanded understandings of both the world and our representations of it as a rich, variegated, interdependent fabric of many levels and kinds of explanations that are integrated with one another to ground effective prediction and action.

2

COMPLEXITIES OF ORGANIZATION

HOW WE THINK ABOUT THE WORLD

How are complex structures different from simple ones? What varieties of "complexity" exist in the natural world? What difference does the scientific study of what are best identified as "complex systems" make in our understanding of nature and the methods by which we come to know it? These are the central questions of this book. My examples come primarily from biology, but the message drawn from these examples applies more broadly. There is clear application to the social sciences whose complexity includes what is true of the biological and goes beyond (see Eve, Horsfall, and Lee 1997; Byrne 1998; Giarratani, Gruver, and Jackson 2007; Streeck 2009; Mantzavinos 2009).

Complexity in biological systems is ubiquitous. It can be seen in the organic composition of a human individual, made up of 46 chromosomes, approximately 210 different cell types, and more than 30,000 genes. A honeybee colony consists of tens of thousands of individuals engaged in a complicated division of labor, with individual task performance responding to internal needs and external conditions of the colony, and patterned by the age and experience of individual organisms (Winston 1987). Or consider *Dictyostelium discoideum*, a eukaryotic slime mold, which lives in two radically different states, one as a population of individual single-celled amoebae and the other as a single multicellular organism formed by the aggregation of the amoebae under circumstances of nutrient deprivation (Kessin 2001). Such complicated, coordinated, and responsive biological systems provide a good context to begin explicating how complex systems differ from simple ones and the varieties of complexity that science now studies. All the biological systems mentioned above exhibit what I characterize as multiple kinds of complexity: multilevel organization, multicomponent causal interactions, plasticity in relation to context variation, and evolved contingency.[1] Understanding that complexity is itself variegated in its "kinds" is essential to grounding the expanded epistemology of integrative pluralism. I will make use of this taxonomy of complexity in what follows. First, I will discuss multilevel organization and evolved contingency. In chapter 3 I will turn to the issue of multicomponent causal interactions and plasticity.

Emergence versus Reduction

The hierarchy of multiple levels of organization in biological systems is well known. Take the human body, made up of molecules (DNA, proteins); cells (neurons or skin cells); tissues (epithelial tissue in the brain); organs (the heart); organ systems (respiration); an individual organism (you or me). What happens to an individual organism happens to its parts, or at least some of them, at the same time. But how does the behavior of the parts determine or explain the behavior of their immediate containing structure or the other ones further up the hierarchy? The traditional reductive strategies of decomposition and analysis "reduce" a complex system to its parts and study the component structures in isolation to understand the behavior of the whole. But, as we will see below, there are different ways in which a system is built out of its parts: from simple aggregation to dynamic feedback structures. Therefore the stories told in the language of the components will have different messages for understanding the behavior of the whole (Simon 1969; Bechtel and Richardson 1993; Wimsatt 1986, 2000, 2007; Kauffman 1984, 1993). The composition of systems from parts varies from context to context; our representations and analyses must be adequate to capture that diversity.

Understanding the varieties of compositional complexity studied by contemporary sciences and describing how complex structures are built out of their component parts challenges some well-worn assumptions about reductive strategies for explanation. Reduction has often been identified as the goal of scientific explanation, but there have been prolonged debates on exactly what reduction means and what kind of explanations it provides (for the classic analysis from a logical perspective, see Nagel 1961; for a more recent treatment, see Sarkar 1998). There are weaker and stronger versions of reduction: ontological, epistemological, and methodological versions (Schaffner 2002, forthcoming). They all share the view that explanation flows "upward" from the behavior of fundamental components to the behavior of the containing system. In its strongest versions, reductionism claims that all causal power resides at a single, fundamental level, so there is nothing to be added to an explanation of system behavior by appeal to higher-level properties. For a simple mechanical example, this picture works. The behavior of a car, though perhaps not its aesthetic appeal, can be reduced to the many behaviors of the parts of which it is constructed. Methodologically, a reductionist strategy advocates the search for explanations for the behavior of some entity in terms of the behavior of its more fundamental components. This strategy has yielded some remarkable advances. Regarding theory reduction, thermodynamic theories of heat and temperature and statistical mechanics are held out by some as a model of successful reduction, explaining macroscopic phenomena of temperature and pressure of gases as "nothing more than" the motions of their molecular constituents.

I will argue, as others have before (Humphreys 1997; Wimsatt 2000, 2007; Bedau 2003; Batterman 2002; Chang 2004), that reduction as entailing a "nothing more than" account of what is causally significant, or what is "real," or what is explanatorily sufficient, fails to capture important features of the behavior of complex systems. Rather than defending a comprehensive antireductionism as the antidote to the seemingly obligatory reductionism, I offer a pluralistic view of valuable explanatory strategies employed by contemporary science, which includes reduction. However, I will argue that, in its strongest form, reductionism alone provides too meager a conceptual framework to capture interesting results of the scientific conceptions and representations of complex systems. It is not a wrong-headed strategy; it is an incomplete one. It should be part of a more full-textured epistemology, not the only game in town.

There are compelling arguments for reductionism. An intuitive picture supports a reductive view of nature and the sciences that describe and explain nature. We live in one world. At least since the rejection of Cartesian mind/body dualism, there has been general agreement that the material from which entities in our world are built is ultimately one kind of "stuff," matter or mass-energy. Even though scientists study different aspects of the one world, fundamental particles, chemical reactions, biological development, cosmological evolution, and so forth, the fact is that they are all studying the *same* world. Thus there is a prima facie reason to believe that the representations and explanations that science develops will stand in some kind of strong relationships to each other. One might argue that there should be a strong coherence and consistency among all good scientific representations and explanations, because, in the final analysis, should we ever approach it, accurate representations capture the relations and causal structures of one world.[2] On this view, alternative true representations of the same physical system are expected to be at least isomorphic to one another for the same reason.[3]

The reasonableness of the assumption that science will succeed through reductive explanations, I believe, rests in large degree on the material composition assumption that every object is made up of only one type of substance, matter or mass-energy. Hence complexity is always the product of a composition of material parts. However, reductionism actually makes a stronger claim, namely, that there is some basic level of *description* that corresponds to the basic fundamental level of matter (see Moser and Trout 1995). Compositional materialism, which motivates reductive strategies, is confused with the existence of a privileged level of description in which all levels of complex structure and behavior can be restated and thus reduced. This assumption fails to accommodate the partial character of any and all representations. Later I will discuss how even Jaegwon Kim's philosophically elegant rejection of emergent properties fails by virtue of neglecting this fundamental fact about representations.

In addition, basing the inevitability of reductive explanations on compositional materialism ignores the ways in which dynamics shape the character of natural systems. Assumptions about causation based on oversimplified compositional materialism, while capturing the simplest forms of causal relationship, ignore chaotic determinism and interactive feedback. I will argue that understanding both the nature of representations and the complexity of causal dynamics preclude a simple-minded reductionism.

The concept of "emergence" is one that poses itself in direct opposition to reduction. Aristotle has been attributed with saying, "The whole is more than the sum of its parts," from his discussion of part-whole causation in the *Metaphysics* (Annas 1976). Since then, philosophers have worried much about what "is more than" means and what "sum" means. There are different senses of emergence that counter the different senses of reduction (see Silberstein 2002; Delehanty 2005). Doctrines of metaphysical emergence that posit a wholly different and unique substance above and beyond the physical are antithetical to physicalism or materialism, an assumption basic to all contemporary science. Although there are those who criticize materialism in the name of theological renderings of the world (Beilby 2002), it has remained a constitutive assumption of science since the seventeenth century that explanation can appeal only to the properties and behaviors of material substance. I agree with this assumption. There are interpretations of emergence that do not conflict with materialism, yet deny the sufficiency of a simple reductive strategy for scientific explanation. These are the only accounts that are meaningful to consider in understanding the complexity studied by the modern sciences.

I maintain that the reductionist presumption that all compositionally complex structures and systems can be explained, without remainder, by appeal only to the properties of their simplest components cannot be sustained. In the nineteenth century British philosophers defined emergence as that which is strictly nonreducible and *therefore* nonexplainable by means of the laws governing its component parts. By their account, espoused by John Stuart Mill (1843), C. D. Broad (1925), and others, emergence is identified with the epistemic marks of nonexplainability and nonreducibility (see McLaughlin 1992). For Mill, qualitative features of water (fluidity, wetness, turbulence) were deemed emergent because there were no explanations of them in terms of the constituent molecules of oxygen and hydrogen. The atomic elements could not be said to be fluid or wet, or to exhibit turbulence. Similarly, emergent biological properties were taken to be inaccessible to explanations by chemical properties, psychological properties inaccessible to biological, and so forth. The nineteenth-century "emergentists" equated explanation with reduction. For them, properties that could not be explained by the causal interactions of constituent parts were emergent. Their properties could not be inferred but only observed, and regularities in and among emergent properties would be built from observations at the higher level, not from laws governing the properties of their component, lower-level parts. On this view, once putatively emergent properties were explained by a reductive strategy, they would cease to be considered emergent.

In fact, twentieth-century science succeeded in providing what are arguably successful reductive explanations of what thinkers like Mill and Broad identified as emergent properties. With the advent of both quantum mechanical accounts of chemical bonding and the explanations of biological phenomena of inheritance by the biochemistry of DNA (see Schaffner 1993), the properties of chemical compounds, like water, and many features of living organisms have arguably been explained in terms of the properties of their constituent parts.[4] The standard examples of emergence appealed to in the nineteenth and early twentieth century were undermined. As a result, there was almost no use of the term "emergence" in science from the 1920s until the 1960s.[5] Bertrand Russell (1927, 285–86) claimed that emergent qualities were merely epiphenomena and of no scientific significance, saying that analysis "enables us to arrive at a structure such that the properties of the complex can be inferred from those of the parts." Talk of emergence died out.

The revival in the 1970s of the use of the term "emergence" in scientific literature coincided with renewed interest in chemical and neurological complexity (see Sperry 1969, 1991; Campbell 1974). It became widespread with the rise of what is now known as complexity science (Prigogine 1997; Bak 1996; Amaral and Ottino 2004; Lewin 1992). In response to the increased scientific interest in emergence, philosophers have developed a number of different accounts of what it is to be emergent. Most of these have discussed emergence of consciousness and the reducibility or nonreducibility of mental states like beliefs and desires to neurological, material properties of the brain (Anderson et al. 2000; Beckermann, Flohr, and Kim 1992; Blitz 1992; Clayton and Davies 2006; Humphreys 1997; O'Connor 1994; O'Connor and Wong 2005; Rueger 2000; Silberstein and McGeever 1999; Stephan 1997). Key features of emergence for both philosophical treatments and scientific applications are novelty, unpredictability, and the causal efficacy of emergent properties or structures, sometimes referred to as downward causation.[6]

A puzzle arises in light of the two tracks—one in philosophy and one in science—of the development of ideas about emergence over the last two centuries. On one hand, philosophical analyses have typically followed Bertrand Russell's dismissal of the explanatory, causal, and scientific role of so-called emergent properties. On the other, there has been an extraordinary scientific revival of interest in emergence. As a crude measure of the latter, an Internet search for the combined terms "emergence," "properties," and "science" in Google Scholar, as of this writing, yields more than 500,000 hits, many of which appear on a cursory inspection to be about emergent properties in natural systems discussed in scientific studies. If the philosophical analyses that dismiss the reality of emergent properties are correct, then why have descriptions of emergent properties in science become so widespread? In what follows I will argue that there are both faulty assumptions and an impoverished conceptual framework that prevents the character of emergent properties referenced by science to be adequately represented in some forms of philosophical analysis. One thing both philosophers of science and scientists agree on: there is only one type of substance—matter—that is constituent of the entities and properties that science engages. If there are emergent properties, they have to be made of the same material substance as nonemergent properties.

Kim's Analysis

I will consider in some detail arguments of Jaegwon Kim, a contemporary philosopher whose views on emergence have set a high standard for clarity in the philosophical community. For those unaccustomed to this style of argumentation, Kim's logic may be somewhat challenging to follow. However, I believe that Kim's philosophical argument against emergence instructively reveals the role of common assumptions among those who reject the attribution of emergence to the kinds of properties that scientists commonly denote as emergent. In addition, I will show there is a structural problem with Kim's strategy of argumentation.

In his 1999 article on emergence, Kim attempts to explicate what emergence could be for a physicalist or materialist. He suggests the following formulation: "Complex systems aggregated out of these material particles begin to exhibit genuinely novel properties that are irreducible to, and neither predictable nor explainable in terms of, the properties of their constituents" (4). The questions Kim is posing are clear. First, how can a higher-level property be grounded in its physical constituents (e.g., water is H_2O) yet be both novel and *not predictable* from the properties of those constituents alone? Second, even if sense is made of such an attribute, how could an emergent property not only be causally autonomous from its physical basis, but actually causally change the properties of its physical constituents? As Kim puts it, "How is it possible for the whole to causally affect its constituent parts on which its very existence and nature depend? If causation or determination is transitive, doesn't this ultimately imply a kind of self-causation, or self-determination—an apparent absurdity?" (28).

Kim uses a standard example from the literature of a phenomenon that allegedly can't be explained by reduction: pain. The logic of his argument is clear. To be emergent is to be nonreducible. But then, what is it for a property to be reduced to its physical or material constituents? Kim suggests a functional strategy (see figure 1). He argues that a higher-order property, like pain, is reducible to physical components if one can describe the higher property functionally and then show the identity of what is referenced by that function with something at the lower, material level. This approach avoids a host of problems of earlier discussions of emergence and reduction that required full-fledged theories at each level of property in order to determine if reduction via logical derivation was successful. A functional description of a property is one in terms of causes and consequences instead of structural components. Take a straightforward example: A chair is describable as an artifact designed to function as something suitable for humans to sit on. It is also the case that chairs are always made out of something material, wood, iron, plastic, and so forth. So a given chair can be described in terms of the material components. Chairs in general are described by their capacity to function as something humans use for sitting. Chair function is *realized by* an individual chair's material components and their structure. The material components of the chair and its structure explain how it has its functional properties. Kim applies the same kind of reasoning to the property of "being in pain." Pain is "being in some state . . . caused by tissue damage and causing winces and groans" (13). Kim *assumes* that whatever stands in the place of that higher-level property—for example, whatever the term "pain" picks out—is also, at the same time, describable in terms of some physical constituents, namely, some material neurological state. The same functional story applies to pain and a particular neural state as it does to the function of the chair and the material chair. Pain is that which is caused by tissue damage and which in turn causes winces and groans. Pain is reducible if it is *nothing more than* what is also referred to by its neurological realization.

Kim argues that, on this account of reduction, for any property to qualify as emergent is extremely difficult. If a higher-level property is just whatever it is that is picked out functionally as what is caused by X (e.g., tissue damage) and in turn causes Y (e.g., winces and groans), then there will always be some configuration of material components at a lower level that can be identified as realizing the functional property of being caused by X and in turn causing Y.[7] To reach this conclusion, however, Kim makes a strong assumption that "every material object has a unique complete microstructural description" (6). If one can translate the functional property at the higher system level into the appropriate part of the *unique, complete* lower-level description, then any explanation can be run at both levels, every prediction could in principle be generated at both levels (whether we currently have the means to do it or not), and nothing at the higher level would be novel or unpredictable given the microlevel material description.

What room is left for emergence on Kim's functional-replacement account of reduction? Not much. To really be unexplainable, novel, and unpredictable from lower-level constituents, a higher-level property would have to be in some sense intrinsic, and not describable in extrinsic,

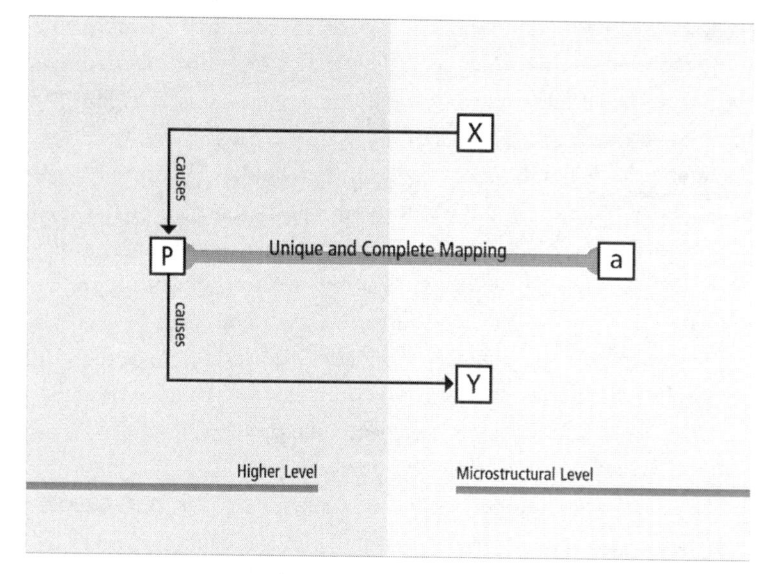

Figure 1. Functional mapping

material terms of what causes it and what it in turn causes. Stephen Yablo (1999, 479) provides a good account of the difference: "You know what an intrinsic property is: it's a property that a thing has (or lacks) regardless of what may be going on outside of itself."

But *scientific* access to properties and entities is entirely by means of what causes them to occur or change and what they in turn cause to occur or change. Anything measurable, for example, could be redescribed functionally in the manner prescribed by Kim. Anything that is functionally describable could then be associated with a lower-level "unique and complete" description. Thus explanations and predictions based on higher-level properties lose their claim to emergence. None of the currently scientifically identified emergent properties (e.g., color patterns on mammals, flocking behavior of birds, division of labor in social insects, etc.) can qualify as emergent on Kim's account. The only plausible candidates for emergence that Kim acknowledges are subjective properties of consciousness, for example, what it feels like to be in pain, not the causes and consequences of being in pain. Such subjective feelings are, as such, not within the purview of scientific study.

Kim gives what he takes to be a clear account of the conditions a property would have to meet to be genuinely novel, unpredictable from the properties of material constituents, unexplainable, and irreducible. For Kim, "if emergent properties exist, they are causally and hence explanatorily, inert and therefore largely useless for the purpose of causal/explanatory theories. If these considerations are correct, higher-level properties can serve as causes in downward causal relations only if they are reducible to lower-level properties. The paradox is that if they are so reducible, they are not really 'higher-level' any longer" (33).

For Kim there is nothing independent of the lower level to be found at the higher level that is even a candidate for causing the lower-level behavior. Thus on his account, nothing available for scientific study would count as emergent. It is not therefore surprising that the main interesting feature of emergent properties in contemporary science, their autonomous causal capacity, turns out also to be impossible to defend if one accepts Kim's analysis. It's a neat logical move. However, it is based on a strong assumption that, I believe, should be rejected. Kim's analysis does nothing to illuminate the realities with which the sciences of complex phenomena confront the philosophical community. I reject Kim's strong assumption that reducibility is a necessary consequence of there being a material realization of higher-level causal relations and will defend an alternative view, a broader notion of emergent properties that makes sense of scientific use and is part of a more complete epistemology.

The problem with Kim's account of reduction and emergence is that, while appearing to be merely preserving materialism, the view that there is no new substance at the higher level that is somehow mysteriously unlike the material substance from which all things are constructed, he actually *imports into the argument a much stronger assumption*, namely, that there is always a unique and complete description of the higher-level phenomena in terms of the lower-level. If we are concerned with *types* of higher-level phenomena (rather than particular instances), then his uniqueness claim is not satisfied. Consider the well-known problem of multiple realizability. Using Kim's example of pain, if we consider pain in general, then each instance of pain will be realized in some neurological microstructure, but that structure may vary radically between instances within and between individuals experiencing the pain. Headaches are one type of pain, but instances of headaches can occur by very different material realizations.

For some, being multiply realized is sufficient to render a property at the higher functional level different in kind from the disjunction of various material configurations that realize it (see note 7). For others, it indicates only that there may be one actual worldly event that can be captured by different vocabularies at different levels of abstraction (Fodor 1974; Putnam 1967; Pylyshyn 1984; Horgan 1993; Sober 1999). But regardless of the conclusion one draws from multiple realizability, there is a shared assumption on how one represents the relationship between higher-level properties and lower-level properties. The common perspective takes a property at a microlevel, compares it to a property at a higher level, and then attempts to determine if there is one real property (perhaps with two descriptions) or two properties that are related by some reductive type of mapping function.

Kim's assumption that any and all phenomena have a unique and complete microstructural description ignores the essential partiality of representation. Any representation—be it linguistic, logical, mathematical, visual, or physical—stands for something else. To be useful, it cannot include every feature in all the glorious detail of the original, or it is just another full-blown instance of the item it represents. Something must be left out, and what is left out is a joint product of the nature of the representing medium (Perini 2005) and the pragmatic purposes the representation serves. A description may be had in microstructural terms, but it will hardly be complete or unique. If one privileges a particular medium, say first-order predicate logic, then one can claim that relative to that system, there may be a unique, complete representation; however, the phenomena itself affords much richer, or poorer, representations as well. What now of the mapping between two different descriptions of a phenomenon, one higher level, one microstructural? The partiality of any representation leaves open the possibility that the two representations will simplify the phenomena in incompatible ways. That is, if the higher-level description individuates features in terms of their functional contribution to an observable effect, while the microstructural individuates features in terms of a numerical precision that is theoretically justified, there may be no systematic way to map one onto the other. For example, what is the microstructural representation of audible sound? The physical representation is in terms of longitudinal and transverse waves, and the frequency of a sound varies continuously. Humans cannot detect all frequencies, but are limited to frequencies between about 20 Hz and 20,000 Hz. Other organisms can detect outside the human range. Discriminating among different sounds may be limited by the structure of auditory organs as well as neuron structure. Which grain of representation is useful for sound units is clearly relative to the detectors in the causal system, whether they be human, animal, or mechanical. What may be left out in one characterization may be critical in another. The descriptive framework for ecological investigations may take as its basic categories predator and prey, systematics may take as its categories genera, species, subspecies, and the like, while population genetics may take as its basic categories allele frequencies. There is no a priori reason why these different taxonomies of objects should coincide (Dupré 1993, 2002; Hacking 1993).

What the philosophical arguments that assume unique, complete representations and a privileged level and straightforward mappings from one level to another miss entirely is a question at the center of much scientific concern with emergence, namely, how is the property at the higher level produced, and what are the differences among the many kinds of relationship between higher- and lower-level properties that occur in nature? Logical analyses of the kind Kim and others adopt are often too static and abstract to represent the dynamic and concrete realities that are the immediate concerns of practicing scientists.

If we take a static snapshot of the higher and lower levels, then the dynamics of *how* the higher level is constituted is lost. Contemporary sciences show us that there are processes, often involving negative and positive feedback, that can generate higher-level stable properties, and these processes are not captured by a static mapping. Kim's attempt to clarify the philosophical conception of emergence has stripped it of any scientifically interesting features, and hence it fails to apply to the properties that scientists have identified as emergent, properties like division of labor in social insect colonies, which have different material realizations for different species of ants, bees, and termites, and perhaps the same dynamics of *how* the higher-level properties are generated. The higher-level properties that naturally emerge by means of self-organization then place constraints on the behavior of their constituent parts.[8]

The philosophical problem with the argument that Kim developed is the conflation of compositional materialism (there is one kind of substance from which all things are created) with descriptive fundamentalism (there is a privileged, complete description of the world in terms of fundamental components). Why is this a problem? All descriptions are abstractions or idealizations. They do not stand in a one-to-one mapping relationship with the entirety of the undescribed world. To think that our language (or any human artifact intended for representation, including mathematics and simulation) captures the material world exactly is something that most post-Kantian philosophers have rejected as simply misconceived. Descriptions are always partial. The metaphysical claim that, at the physical level, there are unique processes that bring about physical results is inescapable, unless one is either a dualist or believes there are uncaused events. However, *representing* these processes in a language, whether that is the vocabulary and syntax of formal logic or of fundamental physics, is an entirely different matter.

It is likely that all the factors contributing to the complete cause of some physical event, say a window breaking when hit by a rock, cannot be represented by any single theory in the syntax of logic or even the language of physics. The local, contingent components of every causal process are just not included in the scope of physical theory or its abstract language (see Cartwright 1994; Cartwright et al. 1995), and these will always be part of the complete cause. There may well be a complete causal process engaged in by *physical* entities: what else could there be? But at the same time there will not be a representation that completely captures this process in terms of *physics* entities. Thus Kim's functional strategy does not work. If there is something in the world that can be isolated by the functional description (caused by X and causing Y), there is no reason to think that a physical description of that piece of the world, partial as it is, will be identical with a higher-level description of that piece of the world, partial as it is. Without the requirement of a unique, complete description at the lower level, the functional identification will not go through, and, as I have argued, there are compelling reasons to reject the assumption of completeness of our representations.

Furthermore, although all properties, events, and structures are physical (what else could they be?[9]), not all physical properties, events, and structures are biological. Compare a conglomeration of molecules constituting a rock with an organization of molecules making up a baby monkey. What's the difference? It is not to be found by looking at what they share, namely, a materialist composition, but only by looking at how they differ. That difference, in part, is what is being picked out by identifying some higher-level structures as emergent. There is an important, explanatory difference that cannot be captured by the emphasis on the reasonable claim that any simple or complex whole is made up of simpler, material parts.

Scientific Emergence

A new understanding of emergence has become widespread in scientific accounts of biological and social phenomena that does not share either the strong epistemic assumptions of the philosophical view Kim advocates or a nonmaterialist metaphysics; that is, it does not preclude explanation from below or above. On this understanding, emergence is identified with certain types of nonaggregative compositional structures, including self-organization (see Wimsatt 1986, 2000; Bedau 1997; Kauffman 1993, 1995; Camazine et al. 2001). Aggregativity is a particularly simple kind of compositional relationship between component parts and the whole. The weight of a pile of rocks being the aggregate of the weight of each component rock is a simple example. For a more complex example, consider the behavior of a car as an aggregate of the behavior of its parts. In these cases, the properties of the whole are predictable, explainable, and reducible to the properties of the components. However, there are a number of ways in which composition can fail to be aggregate and hence fail to be reducible, on Kim's or any other account of reduction.

One way is found in the complexity of chaotic systems represented by nonlinear dynamics.[10] Poincaré in the early 1900s discovered dynamic instability in which a physical system could end up in wildly different end states depending on very small differences in its initial state. This sensitivity to initial conditions is often identified with chaos. Nonlinearity, on the other hand, refers to a behavior that cannot be modeled by a linear equation; that is, it cannot be solved by treating the variables in it as the sum of independent contributions. Feedback refers to a particular way in which processes in a complex system interact, which may lead to chaotic outcomes, adaptive outcomes, equilibrium states, and so forth. Many natural phenomena display these various aspects of complexity.

In December 2006 I visited Rome for a week's holiday. Standing on the balcony of our apartment and looking toward the Pantheon, I observed an intriguing phenomenon. Birds, which I later learned were starlings, were flying in a seemingly dynamically ordered, undulating, spiraling display (see figure 2; for other images, see Rosen 2007; Feder 2007). An early twentieth-century biologist, Edmund Selous, wrote in 1905 (141) about this peculiar behavior of the starlings: "they circle; now dense like a polished roof, now disseminated like the meshes of some vast all-heaven-sweeping net . . . wheeling, rending, tearing, darting , crossing, and piercing one another—a madness in the sky." For Selous this behavior was mad, mysterious, and the only explanation he could envision was one that appealed to some form of collective thought. But today this "madness" is understood in terms of individual birds using standard senses of sight, sound, pressure, and odor to respond to their nearest neighbors, forming a network of nonlinear interactions where information can pass from one to another to permit adaptive collective responses to what is going on in the environment. The move from madness to explanation is not easily rendered in the old Newtonian view. It requires a shift to a more generous view of nature. The world is complex, and now science shows us that so too should be our representations of it. Simple additive relations, simple linear equations, while adequate to explain some simple behaviors, will fail to make sense out of much of the complexity that we find in nature. The starlings' patterns, like the V shape of flocking geese, emerges from the simple interactions of the birds whose individual behaviors constitute it (see Couzin 2007). The V pattern that emerges in a flock of geese or the more complex patterns of the starlings is not predictable by an aggregation of behaviors of individual geese in solo flight, but only from the nonaggregative interaction or self-organizing that derives from the local rules of motion plus feedback among the individuals in group flight. Ontologically, there are just material birds; there is no new substance, no director at a higher level choreographing the artistic patterns the flock makes. This type of behavior is emergent. "It is usually not possible to predict how the interactions among a large number of components within a system result in population-level properties. Such systems often exhibit a recursive, non-linear relationship between the individual behavior and collective ('higher-order') properties generated by these interactions; the individual interactions create a larger-scale structure, which influences the behavior of individuals, which changes the higher-order structure, and so on" (Couzin and Krause 2003, 2).

What the new sense of emergence has in common with nineteenth-century views is that interaction among the parts generates properties that none of the individual components possess, and these higher-order properties in turn can have causal efficacy, that is, novelty. What is different is that for the new scientific view of emergence there are concrete accounts of how and why simple rules of interaction among components produce difficult-to-predict emergent behaviors; that is, prediction and explanation may be possible. Self-organization is grounded in multiple forward and reverse feedback loops that can be radically sensitive to initial and evolving boundary conditions. This differs from Kim's explication of emergence in terms of static property descriptions, and an implicit feed-forward linear type of causation was too limited to represent the scientific notion of emergence.

There are a number of further complex causal interactions that attention to emergent dynamics can reveal. If the feedback system stabilizes a property in the face of fluctuating external conditions—for example, the form of the flock remains dense with a certain percentage of individuals in the interior regions—then this stable property or structure can be a target of natural selection. If some birds organize their behavior in such a way and others do not, and if the structure emerging in the first case presents an advantage against larger bird predation, for example, then this higher-level property will exhibit causal saliency. It will be why one population of birds is more adapted to an environment with lots of large predators compared to another. If the individual behaviors that give rise to the higher-level property are heritable, then evolutionary consequences will follow, depending of course on the trade-off of other adaptive considerations and the force of nonselective processes on the future states of the populations.

Rob Page and I (Page and Mitchell 1991, 1998) developed simulations of group living for social insects to investigate what features at the colony level might be the result of self-organization and what features might have varied in the evolutionary past and thus could be candidates for explanation by natural selection. We attributed to our computer bees minimal features, namely, that individual behavior varies in response to different levels of a stimulus and that the stimulus itself was affected by the behavior of the individual. These features are consistent with those found in solitary insects and thus, arguably, in the evolutionary ancestors of social bees. What we found was that a crude form of division of labor was the result of group living itself; indeed it is inescapable when random individuals interact over time with each other. Thus while further refinements of division of labor, like specific caste ratios, may be the result of natural selection in response to differing environments, once a group is formed, some form of division of labor emerges. Fewell and Page (1999; and see Jeanson and Fewell 2008) forced normally solitary ants to interact and experimentally showed the same result, that is, the emergence of division of labor. The moral is that emergence of higher-level properties from interaction among components via self-organization is not antithetical to features of that property being tuned by natural selection. Rather the two types of processes may work collaboratively to generate the forms and features that we find in evolved complex systems. Selection operating on heritable variation of the components may reductively explain some of the features of higher-level properties like division of labor, and self-organization of emergence may explain other features of the higher-level properties.

Kim's aim was to give an explication of what it is to be emergent in general. The limitations I have outlined show that his analysis fails to cover the typical examples of emergence in the contemporary scientific literature.

There are significant consequences when we expand our notion of causality to include the types of complex interactions that are common for biological systems. The conditions of emergence that appeared as

Figure 2. Starlings over Rome

unrealizable are more easily met. Determinism no longer entails predictability. Even if a behavior, described at a higher level of organization, is determined by the interactions of entities at a lower level of organization, if the dynamics are chaotic or nonlinear, the behavior will not be predictable. The first aspect of emergence identified epistemically with unpredictability or ontologically as novelty automatically assigns the label of emergence to some behavioral outcomes of nonlinear systems.

The nonlinearity that constitutes dynamic complexity of some systems carries with it methodological consequences for understanding such systems. Such behaviors, while deterministic, are unpredictable given their extreme sensitivity to immeasurably small variations in initial and evolving boundary conditions. Many biological systems display features of dynamic complexity including bifurcation, amplification, and a type of phase change (Nicolis and Prigogine 1989; Benincá et al. 2008). Systems exhibiting these behaviors are known variously as "complex," "chaotic," or "nonlinear" systems. A bifurcation is a period doubling, quadrupling, and so forth. It marks the sudden appearance of a qualitatively different solution for a nonlinear system as some parameter is varied. Amplification refers to the nonlinear dramatic increase in response to small increases in the value of some parameters (see May and Oster 1976). Two points in a system may move in vastly different trajectories, even if the difference in their initial configurations is very small. Examples include weather phenomena, fluid turbulence, and crystal growth. Edward Lorenz (1996), a meteorologist, metaphorically termed this type of complexity "the butterfly effect" from the possibility that the flap of a butterfly's wings in Brazil might cause a tornado in Texas. In the slime mold *Dictyostelium discoideum*, the transformation from single-cell populations to multicellular organism is by means of cell-cell signaling and chaotic feedback mechanisms that induce a dramatic macroscopic phase change in the system. Positive feedback can create a situation in which the cellular response changes abruptly and irreversibly once the signal magnitude crosses a critical threshold (Tyson, Chen, and Novak 2003; Goldbeter 1997; Goldbeter et al. 2001; Kessin 2001; Mitchell 2003).

Simulations in addition to laboratory or field experiments are needed to explore the space of outcomes that a chaotic system can visit. The differential equation models that work so well for Newtonian systems are completely inadequate for chaotic systems. Mark Bedau (1997) goes so far as to make the need for simulation to represent system behavior a feature of emergence. That a system ends up in one state rather than another has to do with the immeasurably small differences in the initial conditions under which the very same deterministic rules are applied. It is still the case that the properties and behavior of the component parts cause the ensuing behavior of the system, but there is a shift of emphasis to the features of the history and context that the system experiences to understand why one outcome occurred rather than another. Knowing just the function that describes the causal structure of the parts and our most precise account of its initial state will not tell you in what later state the system will find itself.

The type of emergence that is found in contingently evolved complex systems is both dynamic and extremely context sensitive. Ordered higher-level properties in a complex system emerge and are stabilized by means of feedback and self-organization. Therefore, the history, the context, and the dynamics of systems play leading roles in the explanatory story told by the epistemology of integrative pluralism. These features were relegated to bit parts, typically identified as the source of calculable "boundary conditions" and removable "perturbations," in the chorus line in the traditional epistemology. In many explanations of complex phenomena, the context, the history, and the dynamics occupy center stage.

Unlike some well-known cases of physical causation, in complex systems more than a single cause or a few dominant causes are responsible for the behavior we wish to explain. A billiard ball moves in the direction and with the velocity it does because of the impact on it from the cue ball. Of course there are slight perturbations in the trajectory, due to spin and friction, that would be expected if only the single impact was operating, but for the most part, such behavior is explained by a single dominant cause. Not so in the world of the complex. Of course, as I have already suggested, there are multiple kinds of "complexity." Some are closer to the billiard ball, just with a greater number of causal influences. The multiplicity of factors is not particularly problematic, especially if there are simple rules of interaction, such as additivity. The set of forces on an airplane (a more complicated version of the billiard ball) are identified as the causes of its trajectory and can be summed to predict where it is going. However, complex systems often involve feedback mechanisms resulting in amplification or damping of the results or of nonlinear chaotic behavior, and under these conditions, causal explanations by additivity will fail.

Our understanding of causation expands when we pay close attention to the way in which negative and positive feedback both stabilize phenomena at a higher level and constrain the behavior of the components at the lower level. Feedback provides an operational understanding of one type of downward causation where system-level properties constrain and direct the behavior of the components. Thus a second condition for emergence is met by the complexity studied by contemporary science: there is causal influence of higher-level properties on lower-level behaviors. Feedback loops are operating in these cases. What is new is the interaction of clearly individual behaviors and emergent, higher-level properties of the system as a whole. An example from division of labor in social insects will illustrate.

What causes a honeybee to forage for nectar at a given time, rather than rest comfortably in the hive? Genetic differences among individuals account for some of the differences in foraging frequency. Although the specific pathways are unknown, the correlation between genes and foraging frequency is robust and, in the absence of other interacting factors, explains differences within and between colonies. However, it is also known that individuals change their foraging frequency as a result of environmental stimuli. In a genetically homogeneous population, variation in foraging frequency would depend on the different environmental factors encountered by individuals. In ideal situations, where only one factor is active, a systematic study can reveal the strength of genes or the strength of an environmental stimulus in producing the effects. In natural settings both are operating simultaneously. The result of their joint operation is not a simple, aggregative, linear compound of their individual contributions.

In self-organized systems, feedback interactions among simple behaviors of individual components of a system produce what appears to be an organized group-level effect. Honeybees collect nectar from flowers, which is then processed by digestive enzymes and evaporation into the honey that is found in the combs in the hive. It is the major source of carbohydrates for worker bees and is part of the nutritional substance, along with pollen, that is fed to larvae. Older bees form the caste that flies out of the hive to collect nectar, pollen, and water. An individual forager sucks up the nectar from flowers through her proboscis, and returning to the hive, she unloads it to a younger worker, whose job is to store the nectar in an empty cell. For an individual honeybee forager, the probability that she will continue to forage for nectar may be affected by how long she waits to unload what she has collected upon returning to the hive. The outcome of each individual bee experiencing waiting time collectively generates a system that "tunes" the number of foragers to the "need for nectar" in the hive (Seeley 1989; Seeley and Tovey 1994).

The mechanism by which the system-level property—how much nectar is stored—controls the behavior of individual foragers does not involve any mysterious forces or substances beyond the material makeup of the bees and nectar and hive. But clearly the amount of nectar stored in a hive is not a property of any of the individual bees, although it is the sum of the results of their individual behavior. What is significant for understanding the complexity of this system is to see how an adaptive structure at the higher level (nectar supply) emerges from individual behaviors *and* how the higher-level structure *causally influences* in a feedback loop behavior at the lower level. Clearly, if there is already a lot of nectar stored, then it is more adaptive for foragers to stop collecting, and if there is not much nectar stored, foragers should go out and collect more. Whether a bee continues to forage or begins to rest is caused by the surplus or surfeit of nectar.

When a forager lands with her crop full, she must have the nectar unloaded by another worker bee, who carries it to an empty cell to be stored. The bee who unloads then returns to the next forager waiting to be unloaded. If many of the cells in the hive are already full, it will take the unloading bee longer to find an empty cell into which to deposit the nectar from the first bee, and correlatively if there are many empty cells, it will be quicker. How long a bee waits to be unloaded is a measure of the amount of nectar already stored, and this waiting time is a trigger for continuing to forage or stopping.

Self-organization and feedback make scientific sense of emergent features of complex systems (see Camazine et al. 2001; Bonabeau et al. 1997). Higher-level properties—the pattern of division of labor, the amount of nectar stored in the hive—are caused by the interactions of the components, but not in a simple aggregative way. Interactions are often chaotic, displaying both positive and negative feedback that can generate novelty in the overall response that is not predictable from the intrinsic properties of the individual components. And in turn, the higher-level properties can influence the behavior of the components whose actions themselves determine the higher-level properties, as is the case in nectar-foraging behavior and amount of nectar stored. The "self-causation" in a system that Kim had alluded to as an "apparent absurdity" turns out to be a common feature of the type of complex systems investigated and explained by modern science. What is required to make sense of scientific emergence is a richer conceptual framework that accommodates the dynamics by which the behavior of the component parts of a complex system generate a higher-level property as well as how that higher-level property in turn can causally influence the very components from which it has been generated. Not all higher-level properties are emergent, but some are. Taking account of nonlinear dynamics and feedback causal processes in addition to static and linear representations that may be adequate for simpler domains of nature is one of the ingredients of an integrative pluralistic ontology and epistemology.

3

COMPLEXITIES OF EVOLVED DIVERSITY

LAWS

Evolved diversity is another important aspect of the overall complexity of the natural world. Evolved creatures display varied, alternative adaptive solutions to similar selective regimes, and the same adaptation may be reached by very different means. Evolved features of complex systems record the historical contingency, variability, and path dependence of the particular array of properties and behaviors that are found in biological systems. Variability is endemic to biological populations; indeed it is a necessary component to the evolutionary dynamics that generate adaptive change. Given certain aspects of speciation processes (e.g., niche specialization), variation both in distribution of traits within a population and in the traits characteristic of different populations is fundamental to evolved complex biological systems. The operation of natural section, which limits diversity in the face of multiple forces driving greater and greater variation, makes the biological world complex in ways that not only involve rich feedback loops but also depend on history and context.

Consider the Galápagos finches whose diversity Darwin initially thought was indicative of their being as unrelated as blackbirds are to wrens. After Darwin returned to London from collecting the birds on the voyage of the *Beagle*, the leading ornithologist of the day, James Gould, determined that the various birds were all species of finch. The variation in beak size and shape, from one as large as a grosbeak to one as small as a warbler with gradations in between, was a result of niche specialization (see Grant and Grant 1986). The different food sources on the different parts of the islands where the finches were found correlated with the beak size and shape. These variations show how from a single shared ancestral mainland finch, diversity could be generated. As Darwin says (1839, chapter 17), "Seeing this gradation and diversity of structure in one small, intimately related group of birds, one might really fancy that from an original paucity of birds in this archipelago, one species had been taken and modified for different ends."

Leo Buss (1987) has documented the great variety of organization of multicelled life on our planet and argued that multicellular organisms may have evolved, in part, in the process of niche occupation. New, more complex forms of life could achieve novel abilities in mobility and feeding that allow them to access new niches inaccessible to less mobile organisms. Relocating in a new niche would take the multicellular forms out of the competitive game with single-celled organisms. While the process of natural selection continually limits existing variation by preserving only the relatively most fit, individual variation is continually renewed by mutation and recombination. And variation, as the finches and the history of multicellularity show, is an outcome of speciation and major transitions in evolution (see Maynard Smith and Szathmáry 1997).

There is diversity not just in outcome, but also in the means to attaining adaptations. The case can be made in terms of both ultimate evolutionary pathways and proximate mechanisms for generating a complex system phenotype. Division of labor has evolved independently from solitary ancestors multiple times in ants, bees, and wasps.[1] This colony phenotype consists of different workers specializing in different parts of the entire set of behaviors being performed in a colony. In each species, the queen produces offspring while the workers are sterile, and different individual workers vary in tending the brood, cleaning the cells, guarding the entrance, and foraging. Some of the variation is a function of the age of the worker (age polyethism), some a function of physiological differences (morphological polyethism). While there are thought to be ergonomic benefits to sharing labor in a colony compared to solitary life (Oster and Wilson 1979), the evolutionary paths to achieving this adaptive result are likely to be somewhat different. One reason is that the genetic relatedness among workers in a colony varies among species.

Take, for example, honeybees and fire ants. Both display division of labor in reproduction, where the queen produces offspring while the workers are sterile, and in provisioning and maintaining the colony, with individuals of different ages performing different tasks: tending the brood, cleaning the cells, guarding the entrance, and foraging. However, how they achieve the same complex structure differs. Honeybees harbor much more genetic variation than do fire ants (Boomsma and Sundström 1998). Honeybee queens mate with up to seventeen different drones on their nuptial flight. Thus the female worker offspring in a bee colony can have seventeen different fathers, resulting in complicated and varied degrees of relatedness among the workers. The black imported fire ant queen mates just once; thus all the worker offspring are full sisters. That social insects share a peculiar haploid-diploid genetic structure has been used to explain the division of reproductive labor between queens and workers. But while some shared features may be explained by this similarity, the means by which different groups achieve this outcome may vary.

Hamilton (1964), in formulating his theory of kin selection, offered an explanation for the evolution of sterility in social insects, a problem for a Darwinian account of natural selection favoring traits that increase individual reproductive success. Hymenoptera (ants, bees, and wasps) have an unusual genetic structure, where males are haploid and females are diploid. That means that males are the result of unfertilized eggs, produced only by the queen and having only half the chromosomes and genes of the female workers. Females receive genes from both the queen and the male who fertilizes the egg from which she comes. Hamilton's elegant argument showed that full sisters in this structure on average share three-quarters of their genes with each other, while mothers and daughters share only half of their genes. Given this, female workers contribute more of their individual genes to the next generation by raising sisters than by having offspring of their own. These conditions explain why female workers are functionally sterile, the foundation of division of labor.

In comparing fire ants with honeybees, it is clear that the more complex relatedness in the latter impedes a straightforward application of Hamilton's argument. Indeed, the evolutionary reasons for selective advantage of division of labor in honeybees may have to do not just with the haploid-diploid genetic structure, but also with parasite resistance and other forms of robustness afforded by maintaining division of labor in conjunction with genetic variability in the colony (see Breed and Page 1989; Page and Metcalf 1982; Waibel et al. 2006). The variability of pathway is required by the differences in genetic resources found in different species of ant and bee.

Different internal and external factors have been modeled as proximate causes of the variant behavior of individuals constitutive of division of labor (see Beshers and Fewell 2001; Gordon 1989). The factors include differences in genes, hormone levels, experience, location in the hive, and access to information. Clearly the role of genetic differences in accounting for differences in individual behaviors will be in part a function of the genetic resources available in the population. Given that honeybees and fire ants harbor different degrees of genetic variability, some proximate mechanisms may be active in one system that are not available to the other to generate division of labor. Different pathways cause a similar colony property that has adaptive advantage for both species in the two complex systems (see also Foitzik and Herbers 2001).

In contrast, there are cases of similar selective environments, which produce similar adaptive solutions from unrelated species, a form of convergent evolution. Consider the similar environmental niches filled by marsupial mammals in Australia and placental mammals in Europe or North America. Kangaroos are marsupials that fill the niche of placental deer, and koalas are marsupial counterparts to placental sloths (see Springer, Kirsch, and Case 1997). While some of the features of the species that fill the niche are similar, others are not. The diversity of life combined with the correlation of features with environmental contexts was a major source of puzzlement and inspiration for Charles Darwin, who saw similarities in form within a location as due in part to common ancestry while similarity in form in different locations was attributed to similar environmental challenges (see Darwin 1859, chapter 14).

The scientific investigation of evolved diversity of the kinds described above challenges a central tenet of traditional views of knowledge: complexity can be explained by reduction to simple laws. It has long been lamented by some that biology has no laws similar to the purported laws of physics (Smart 1963; Beatty 1995). That is, products of evolution and the rules that describe their behavior appear to be more contingent and less universal, more ephemeral, less fundamental, and hence subject to change under a variety of contextual scenarios. The standard is clear, and it is the one that the nineteenth-century philosophers codified and philosophers have discussed ever since: To be a law, a generalization must be universally true, exceptionless, and naturally necessary. Universality means that a law applies to all space and time, and if it is true and universal, then there will be no exceptions. Nothing will escape obeying a law of nature. There is almost no generalization true of the biological world that can meet these criteria of a scientific law.

Natural necessity is a more difficult idea to understand. The *intuition* behind the requirement of natural necessity, however, is clear. Some universal truths seem to be only accidental, while others carry a stronger warrant. Consider the difference between the following two statements: All gold spheres are less than a mile in diameter. All uranium spheres are less than a mile in diameter (Goodman 1947). Both of these are true, there are no exceptions, but the truth of the claim about uranium seems to be necessary while the claim about gold, though true, seems to be accidentally so. It could have been the case that enough of the element of gold was produced that a sphere of gold could be that large. But there is something not accidental about the claim about uranium; its instability precludes having a sphere a mile in diameter. Such a sphere would self-destruct. That difference is what is referenced by appeal to natural necessity.[2]

Generalizations describing the behavior of the biological systems produced by evolutionary processes of random mutation and natural selection do not appear to meet these stringent criteria for lawfulness. There is even doubt that we have discovered any such strict laws in physics (Cartwright 1983, 1994; Earman, Roberts, and Smith 2002). Nearly everyone agrees that what we have discovered about organisms, species, and ecosystems are contingent, local truths.[3]

Philosophers have struggled to reconcile the mismatch between the realities of biology and the strict view of laws, adopting various strategies from appending what are known as *ceteris paribus* clauses to these lesser truths to render them formally universal, to relegating biology to the realm of the lawless (Sober 1997; Brandon 1997; Lange 2000). The stakes are high, as laws are what science is said to search for and are at the core of traditional accounts of explanation (see Hempel and Oppenheim 1948). If biology has no laws, then it risks being judged something less than a science.

Does Biology Have Laws?

How do we decide whether biology has laws? I believe there are three strategies for pursuing this question: a normative, a paradigmatic, and a pragmatic approach (Mitchell 1997, 2000). The normative approach is the most familiar. To proceed, one begins with a norm or *definition* of lawfulness, and then each candidate generalization in biology is reviewed to see if the specified conditions are met. If yes, then there are laws in biology; if no, then there are not laws in biology. The paradigmatic approach begins with a set of *exemplars* of laws (characteristically in physics) and compares these to the generalizations of biology. Again, if biological generalizations are not like the exemplars, then biology is deemed lawless. The pragmatic approach focuses on the *role* of laws in science, and queries biological generalizations to see whether and to what degree they function in that role.

My approach to the problem is to re-envisage lawfulness functionally. That is, rather than try to squeeze biological generalizations into the uncomfortable cloak of universal, exceptionless truths—either by restricting the candidates to mathematical truths (Sober 1997), or by tacking on a ceteris paribus clause (Sober 1997), or by giving up lawfulness altogether (Beatty 1995, 1997; Brandon 1997)—I ask what do laws of the traditional form *do* in science, and open up the question of whether other forms of truths can satisfy the *function* of laws. The answer is yes. Exception-ridden, nonuniversal true generalizations can, under clearly defined conditions, function in the same way that universal, exceptionless generalizations do in explanation and prediction. They are not as easy to use, but they are usable nevertheless.

We can make sense of the fact that the patterns of behavior we see in social insect colonies or the patterns of genetic frequencies we see over time in a population subject to selection are caused, are predictable, are explainable, and are subject to predictable changes we desire as a result of our actions. Yet the kinds of causal knowledge that permit us to perform all these epistemological and pragmatic tasks include statements of regularity other than universal, exceptionless truths. Some are contingent, domain-restricted truths. Some generalizations regarding division of labor and underlying genetic differences will apply to honeybees, but not to fire ants. Inferences about similar adaptive response to similar environmental conditions will sometimes apply, and sometimes not.

My strategy is not to take either side of the debate, that is, not to say, "Yes! Biology and the other special sciences do have laws just like physics (or close enough anyway)" or "No! Biology and the other special sciences can never have laws like those of physics; knowledge of the complex, contingent evolved systems has a wholly different character." Rather I suggest replacing the standard conception of laws that has structured the debate to this point with a more spacious and nuanced conceptual framework that not only illuminates what it is about knowledge of complex biological nature that is similar to knowledge of the physical, but also exposes how they differ. I maintain that our philosophical conception of the character of scientific knowledge should be based on what types of claims function reliably in scientific explanations and predictions (Mitchell 1997, 2000, 2003). Some generalizations are more like the strict model of laws; others are not. Both should be called laws. The more inclusive class of regularities, which I call pragmatic laws, are widespread in biology and, arguably, beyond.

On my revised view of lawfulness, biology has laws, though they are more contingent than some laws of fundamental physics. This is not merely a strategy of redefinition; it has normative force for scientific practice. I believe my approach changes the areas of focus for discussions of how laws are identified, how they are related to evidence, and how they are used to predict and act. There are a number of questions raised by the contingency of the generalizations discovered about biological systems.

- *What* makes a generalization contingent; that is, what does its truth depend on?
- *Why* is a generalization about biological systems more contingent than generalizations of physics or chemistry?
- *How* does the contingency of biological claims factor into our use of them to explain, predict, and intervene in the world?

Contingency versus Natural Necessity

The standard philosophical characterization of laws requires them to be universal in scope, exceptionless, empirically true, and naturally necessary. These characteristics have been identified by scientists and philosophers as rendering laws critical components of scientific explanations. (A classic account of the role of laws in explanation is found in Hempel and Oppenheim 1948.) Newton's law of universal gravitation was the paramount example for the nineteenth-century philosophers. That law says all bodies in the universe attract each other with a force proportional to the masses of the bodies and inversely proportional to the square of the distance between them. Interestingly, this law proved to be empirically inadequate. The law of general relativity, developed by Albert Einstein, now replaces Newton's law. Indeed, general relativity is the better candidate for a genuine law of nature, in terms of exhibiting the properties of the traditional account. Yet physicists still teach and use "Newton's law."

Physicists don't have a corner on the "laws" market: biologists also speak of "laws" in their writings. Mendel's law of segregation states that for any sexually reproducing organism, gamete production preserves a 50:50 ratio of alleles[4] from the organism's two parents. In an oversimplified example, say you have a gene for blue eyes from your father and a gene for brown eyes from your mother and there are only two genes that occupy the "eye-color locus" on a chromosome. Then 50% of the gametes (either eggs or sperm) that you produce will carry the blue-eye gene, and 50% will carry the brown-eye gene. In contrast to the required universality, however, there are exceptions to this law. In some contexts, a mechanism known as segregation distortion, or meiotic drive, skews the genotypic frequencies from the Mendelian expectations (Taylor and Ingvarsson 2003). In such cases, one of the genes will be overrepresented in the population of gametes. The existence of exceptions is typical for generalizations in biology. If laws play a significant role in science for prediction, explanation, and intervention, then the lack of them in biology seemingly raises a serious problem.

How are we to think about the knowledge we have of general features and patterns of biological systems that fail the characterization as universal, exceptionless, necessary truths? Biological sciences are *not* merely collections of accidental truths. Yet what options are there for classifying biological knowledge claims? The failure of biological generalizations to conform to strict lawfulness has been blamed on the *contingency* of evolved complex structures. They lack the natural necessity that laws of physics seem to possess. The ways biological systems are organized have evolved, and thus the generalizations true of the biological world have changed over time with the evolution of new, contingent structures. The causal structures not only could have been different, but in fact *were different* in different periods in the evolution of life on the planet and *are different* for the set of evolved structures at a given time. Thus exceptionless universality and necessity seem to be unattainable. The traditional account of scientific laws is out of reach for biology.

A philosopher of biology, John Beatty, has dubbed the fact that generalizations about evolved biological systems fail to meet the standard requirements for scientific laws "the evolutionary contingency thesis." "To say that biological generalizations are evolutionarily contingent is to say that they are not laws of nature—they do not express any natural necessity; they may be true, but nothing in nature necessitates their truth" (Beatty 1995, 52). As I will argue, the requirements for lawfulness fail to reflect the reality of scientific practice. As a consequence, the traditional understanding of laws is incomplete and fails to account for how humans have knowledge of the complexity of our world.

Stephen Jay Gould in *Wonderful Life* (1990) provided a metaphor for picturing the contingency of biological form and behavior. He suggested that the evolved forms and behaviors that populate the earth in the twenty-first century could well have been different, if only a few small events had been otherwise in the past. He draws an analogy with Frank Capra's 1946 film *It's a Wonderful Life*, where the character of George Bailey, played by Jimmy Stewart, gets to visit the world as it would have been if he had never been born. What George Bailey finds is a rather dreary version of his hometown, Bedford Falls, wracked with crime and poverty, in the place of the more hopeful town that developed under the condition that Bailey actually had lived. The moral is clear: Bailey's existence made all the difference between a dreary and a thriving Bedford Falls. What Gould suggests is that evolution is similarly capricious. If we "roll the tape back" on evolution, like George Bailey (or Frank Capra) did in the story in the film, then what unfolds the second, or third, or more time around would be different in each case. As Gould puts it,

> I call this experiment "replaying life's tape." You press the rewind button and, making sure you thoroughly erase everything that actually happened, go back to any time and place in the past—say, to the seas of the Burgess Shale. Then let the tape run again and see if the repetition looks at all like the original. If each replay strongly resembles life's actual pathway, then we must conclude that what really happened pretty much had to happen. But suppose that the experimental versions all yield sensible results strikingly different from the actual history of life? What could we then say about the predictability of self-conscious intelligence? Or of mammals? Or of life on land? Or simply of multicellular persistence for 600 million difficult years? (48–50)

Gould's questions are rhetorical. Clearly the answer is that few features of a complex organism *had to happen*. There are always considerable degrees of freedom for evolution to move the cast of existing characters in more than one "necessary" direction. I remember when Mike Gilpin, a biology professor at University of California, San Diego, was first introduced to my new dog, Emerson. Emerson is a Cardigan corgi, a breed that is sometimes described as a foot tall and a yard long. Gilpin's only comment was, "What an improbable creature." That comment might be generalized to all species.

Consider the flamingo. These birds have a unique form of life, with an unusual manner of feeding. They stand in shallow water, bend their necks, tilt their bills upside down, and wave their heads sideways. Their tongue shoots in and out, drawing water into the front of the bill and pushing it out the sides while fringed plates on the tongue trap algae and crustaceans. On the face of it, a flamingo is another improbable creature. Are flamingos a necessary consequence of the laws of nature, unfolding in Laplacean lockstep from the big bang until the appearance of their primitive ancestors 50 million years ago?[5] Why do flamingos engage in such peculiar feeding behavior, have pink or red feathers as a result of the carotenoid pigments found in their food, or possess a type of erectile tissue unique among birds (Holliday et al. 2006)? In short, why do these strange creatures exist and have the properties and behaviors they do? If we rolled the tape of evolution back 60 million years and let it play again, would there be flamingos? The judgment of contemporary biology is no.

The chanciness of the existence and features of evolved biological systems seems to set biological objects apart from those of physics or even of chemistry. Physical particles obey laws, and their behavior is explained by those laws wherever and whenever such particles occur. Indeed the Newtonian image of science urges one to search for the fundamental laws that govern the most basic parts of matter with the hope that these will explain all that happens. But the complex evolved structures that constitute the domain of biology (and many other sciences, especially social sciences) do not fit the prescribed pattern.

But, we might well ask, are the laws of physics or of chemistry necessary? The evolutionary contingency that Beatty attributes to biological generalizations does not separate biological generalizations from those of the other sciences.[6] All scientific laws or laws of nature are contingent in two senses. First, they are clearly logically contingent. Second, they are all "evolved" in that the relations described in the law depend upon certain other conditions obtaining. For the version of Galileo's law of free fall that specifies the distance all bodies fall in a given time as equal to

1/2 × 9.8 meters per second per second × the time of fall squared to truly describe relations of bodies in our world requires that the mass of the earth be what it is. If, for example, the core of the earth were lead, which has an atomic weight of 207.2, instead of iron, which has an atomic weight of 55.85, the quantitative acceleration would be four times what Galileo's law specifies (although the force producing that acceleration would still be described by Newton's inverse-square relation). The acceleration due to gravity in Galileo's law—that is, 9.8 meters per second per second—is contingent on the composition of the earth. That the earth is configured the way it is, is the result of the origin of the universe, the creation of the stars and planets, and the resulting matter and energy distributions in our part of our galaxy. Generally stated, there are conditions in our world upon which the truth of laws, like Galileo's law of free fall, depend. They all could have been otherwise. This is the case whether those conditions are the result of particular episodes of biological evolution and are subject to further modification, or whether they are conditions that were fixed in the first three minutes of the birth of the universe. Whatever else one believes, scientific laws describe our world, not a logically necessary world.

All laws are *logically* contingent, yet there is still a difference between Mendel's law of 50:50 segregation and Galileo's law of free fall and Newton's law of universal gravitation. How can we represent that difference? That there is a difference between Mendel's laws and Galileo's and Newton's laws should be explained, but it is not the difference between a claim that could not have been otherwise (a "law" in the traditional sense) and a contingent claim (a "nonlaw"). What is required to represent the difference between these laws is a framework in which to locate different *degrees of stability* of the conditions upon which the relation described is contingent. The conditions upon which the different laws rest vary with respect to their stability in either time or space or both.

The dichotomous opposition between natural contingency and natural necessity of the kind we find in traditional philosophical analyses of laws like Nelson Goodman's and many of those that have followed is a product of framing natural relations in logical terms. In logic all statements are either necessarily true, in virtue of their form or the meaning of their terms, or they are contingent, in that their truth depends on facts.

Mathematical truths, like 2 + 2 = 4, are necessarily true. They could not be otherwise. The claim "All bachelors are unmarried" is necessarily true, due to the meanings of the terms. A formal claim of identity "A is A" is also necessarily true. But factual statements—"There are four apples in the bowl," "John is a bachelor," or "Acceleration is a function of force"—are contingent statements in that they could be either true or false. Contingent truth depends on features of the world about which they make claims. There is no middle ground between *logical* necessity and *logical* contingency. A statement is either one or the other.

The importation of the logical dichotomy to partition empirical scientific generalizations into naturally necessary or contingent fails. Yet I believe this very importation of the logical distinction into science supplies one basis for the traditional view of scientific laws as "naturally necessary." Modeling natural necessity on logical necessity carries with it the presumption that the latter, like the former, is an all-or-nothing property. Logically a statement is either necessary or contingent. Nomologically a relation between two events in the world in the traditional view is taken to be either necessary or contingent (i.e., accidental). But as claimed above, relations between events display varying degrees of contingency and stability. A mistaken dichotomy propagates a mischaracterization of the sciences of complex phenomena, including much of what can be learned about evolved biological structures, as not being "real science" because they describe what is only contingent or accidental.

The difference between generalizations in physics and those in biology or the social sciences is inadequately captured by the dichotomy between necessity and contingency. Any scientific truth describes events that could have been otherwise, whether it is about the physical constituents of our world or the biological ones. Thus the "laws" of physics and the "laws" of biology are both strictly contingent; their truth depends not on logical form or definition, but on whether they accurately represent our world. There are differences, but they are differences in degree and origin, and not in logical kind. The details of the differences demarcate an important domain of study for philosophy of science.

Changing the conditions upon which Galileo's law depends would have downstream consequences of a very different kind than changing the historical conditions upon which Mendel's law depends. The lawful relationship between free-falling bodies and the earth and parent and gamete frequency have different degrees of stability and scope, which affects the degree to which we can depend on them holding in many contexts (see discussion on stability below and Mitchell 2000). The actual acceleration of falling bodies, given the conditions of how the earth is actually configured, is deterministic, while Mendel's law is probabilistic. Thus they differ in the degree of contingency they display because they differ in the stability of the conditions upon which they depend, as well as differing in the strength of relationship, deterministic or probabilistic.

The stability of the conditions upon which a causal relationship depends establishes a continuum, rather than a dichotomously partitioned space of the necessary and the contingent. There is no clear metric to use to measure degrees of stability, but there is, I believe, an ordering among the domains of the sciences that sheds light on the differences in the explanatory problems that face biologists and social scientists compared to physicists (see Mitchell 2000 for a more complete exploration of these issues).

Necessity, Possibility, and the Contingent Universe

What are the differences between generalizations discovered about evolved complex biological systems and the chemical and physical truths that are also true of biological systems? In evolutionary biology what is actualized depends on the history of how things developed on our planet and the composition relations describing how complex systems are built out of their components. Yet not every biological form that is possible has ever existed; some forms may never be actualized. This is true too of the physical and chemical forms and the laws that govern them. Our universe is not the way it is because it is logically necessary that it be this way. In that sense, our particular universe realizes some subset of the logically possible relations and structures that might have come into existence but in fact did not (see figure 3).

The domain of the physically possible contains in it the subset of the physically actual—which I presume does not in fact occupy the space of all of physical possibility. For example, there might have been different

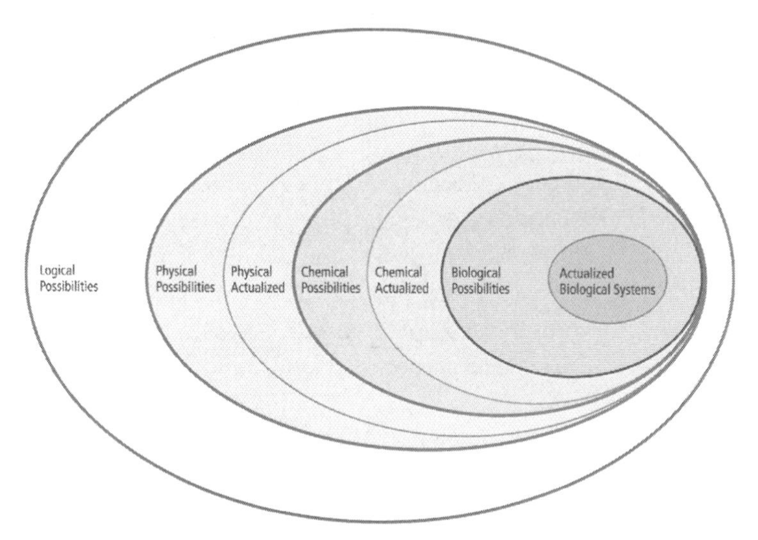

Figure 3. Possibility and actuality

quantities of matter in the primordial atom at the time of the big bang, and this would have had downstream consequences on the types of natural relationships and causal laws that would have obtained (see Smolin 1997). Perhaps what should be meant by physically or naturally necessary is that once some fundamental features are fixed, then the further relationships that can hold among physical properties and events become necessary *contingent on* those historically prior features.

One can then ask, presuming some physical features of our universe have been established—say during the first three minutes after the big bang (Weinberg 1993)—what further physical events would transpire? In the evolution of the stars, could the quantity of elements created vary if "we rolled back the tape" of the physical history of the universe and played it again? What kinds of biological forms would be possible at any given point in the physical history of the universe? What biological kinds have been actualized?

For biological organisms and their behavior, as the evolutionary contingency thesis maintains, at any point in history, life could have gone a different way without any changes in the physical laws that constrain biological form. Chance, say in the form of radioactive decay, can induce mutations in germ-line DNA, thereby introducing new variations into a population under selection that may change the evolutionary trajectory for that population. Since ionizing radiation can create mutations (see Forster et al. 2002) and radioactive decay is probabilistic, this would be one plausible source of the contingency of evolved biological forms. Or consider an unexpected catastrophic event like the impact of an asteroid on the earth. This has been hypothesized as the cause of the mass extinction of plants and animals, notably dinosaurs, approximately 65 million years ago (Alvarez et al. 1980). Had the asteroid missed the earth, what would the population of species and the laws that govern species behavior be like now?

Path dependence generated by natural selection for the relatively more fit variants conjoined with stochastic events, like ionizing radiation or catastrophes, locate the domain of actual biological forms into a specific region of the domain of possible biological forms. What biologists primarily investigate is not what is biologically necessary, given the constraints upon biological form and behavior from the actual physical and chemical relationships available. Rather they are concerned to explain what is contingently present within the wider domain of the biologically possible (Dawkins 1996; Dennett 1995). That is not to say that there is no interest in what is biologically necessary. Fontana and Buss (1994) have developed a model for what all biological organisms must share, that is, features of self-maintenance and self-replication. McShea (2005) argues that there are inevitable tendencies toward some features of complexity for evolved biological structures, and Brandon (2006) proposes a first law of the natural state of biological populations. But while universal features will explain something about the organisms that have occurred in the past and currently populate our planet, it does not explain much of why what is here now does what it does.

This view of the complex natural contingency relationships among the domains of the different sciences, in terms of the constraints that flow upward from features of the components, provides a framework in which to explore what and why knowledge of the parts explains something about the behavior of the containing structures. Physics constrains but does not determine the *actual* chemical properties that occur at any given time in the history of the universe. Consider, for example, the periodic table, which organizes the elements in a series of increasing atomic number that equals the number of protons in the nucleus. While some elements, like gold, have been known since prehistoric times, others were discovered later. Phosphorus was discovered in 1669 by Hennig Brand, and only twelve elements were known before 1700. The theoretical existence of the superactinides, a series of undiscovered chemical elements from atomic numbers 121 (unbiunium) through 153 (unpenttrium), was proposed by Glenn T. Seaborg, winner of the 1951 Nobel Prize in Chemistry. Do they actually exist? They are certainly physically possible, but so far no discovery has been confirmed (Rouvray and King 2004). Chemistry constrains but does not determine the behavior of the evolved biological structures that occur at any given time in the history of life, and so on. A scientist can choose to explore features that are more or less global (and physics in general will always be more global than biology in general). In different types of investigation, and for different types of structures, the role that context plays—that is, the features typically not represented as being part of the object under study—will vary. Biologists are primarily focused on the actual evolved forms, rather than the entire domain of the biologically possible. The causes that drive actual biological form into the region it occupies are evolution by natural selection, developmental constraints, and chance. Evolutionary contingency points to the fact that none of the forms and rules that govern actual biological systems are necessary even given the constraints from "below"; they could have been different even granting the same physical components they are made from, and they may well be different in the future.

> Crank your algorithm of natural selection to your heart's content, and you cannot grind out the contingent patterns built during the earth's geological history. You will get predictable pieces here and there (convergent evolution of wings in flying creatures), but you will also encounter too much randomness from a plethora of sources, too many additional principles from within biological theory, and too many unpredictable impacts from environmental histories beyond biology (including those occasional meteors)—all showing that the theory of natural selection must work in concert with several other principles of change to explain the observed pattern of evolution. (Gould 1997, 47–52)

What follows is that there could be a change in the domain of biological structures without necessarily a change in the underlying physical stuff, if the change was one that was built solely on structure and organization. The underlying biochemistry of a bee, for example, remains the same for solitary and social insects. But the behavior of the solitary and social are different and obey different rules, and that is a result of the social context in which they live and the feedback that develops in that context, not their basic biology. Reductive approaches to explaining the behavior of complex systems get you something, but not everything we want to explain. Focus on universal features of complex systems gets you something, too, but not the contingent outcomes of natural selection and chance.

What then is the relationship between the truths we can glean of the physical universe and those we can come to understand about the biological parts of that universe? Most of the laws of fundamental physics are more universally applicable, that is, are more stable over changes in context, in space and time, than are the causal relations that hold in the biological world. On my view, these are variations in degree—not in kind. Nevertheless, many of the relationships connecting physical properties and events are more stable than are the relationships connecting biological properties and events. What stability denotes is the degree of invariance of a relationship between events or properties that are represented in scientific generalizations. The traditional view of laws required that stability be implacable. The relationship between mass, distance, and gravitational attraction would hold, come what may. But stability varies. Some structures are more stable than others, are less vulnerable to being disrupted by what occurs in their neighborhood, but few, if any, satisfy the strictest conditions of exceptionless universality. There is a difference between fundamental physics and the biological and social sciences—but it is not the difference of a domain of laws versus a domain of accidents. By broadening the conceptual space in which we can locate the truths discovered in the various scientific pursuits, we can come to understand and represent the nature of those differences. Thus the interesting issue for biological knowledge becomes not charting how results can be made more and more like knowledge of fundamental physics, but how to characterize the types of contingent, complex causal dependencies found in that domain. Study of such issues defines research programs both for biology (and other sciences of the complex) and for philosophy of science.

It is not sufficient to say *that* laws are contingent; one must detail *what* kinds of conditions they depend upon and *how* that dependency works (Mitchell 2002b). There are varying degrees of stability of causal contingency, and there are different kinds of contingency. Only by further articulating the differences, rather than relegating biological contingent truths to the same class of accidents as the restrictions on the diameter of gold spheres,[7] can the nature of complex systems be taken seriously. For the problem of laws in science is not just a feature of our epistemological failings; it is a function of the nature of complexity displayed by the objects studied by biology and the social sciences. Providing a more adequate understanding of laws requires a better understanding of contingency so that we can state the many ways in which laws are not always "universal and exceptionless." Only then can we hope to use that knowledge to explain, predict, and intervene in the world.

Biologists and social scientists are providing explanations of causal systems that are historically, evolutionarily, and developmentally contingent. The role of the naturalized philosopher of science is to study these practices, reflect upon the assumptions that make sense of them, and articulate what lies behind successful science. This book argues that the result of such a study is a provocation to revise the philosophical landscape to better capture what science achieves. We need to replace the dichotomy of law versus accident with a more nuanced continuum of contingency to better describe the character of the systems studied. We need to analyze how different factors contribute to contingency (Mitchell 2002b), distinguishing between the ways causal dependence is shaped by, say, chance and adaptation in evolutionary history, or early and late development in ontogeny as well as by features of the context that complex systems occupy. Epistemology is the study of knowledge, and philosophy of science takes the best case of human knowledge to be displayed by science. But science has changed since Newton's grand unification of the laws of motion in the seventeenth century. Philosophers of science, by considering not just what was true of science then, but what has changed since Newton, have new ground to cover. One task ahead is to analyze and articulate the different types of contingency, thereby expanding our conceptual frameworks to include the science of multicomponent, multilevel, evolved complex systems.

Let us return to the initial case of complexity described in this book, the relationship between variant genetic factors and the incidence of major depressive disorder. Is there a universal, exceptionless law that links genes to the psychiatric disorder? The answer is no. There are many different factors in different combinations that can result in the disease. Some of these pathways will involve the short allele for the promoter region of the serotonin transport system neurons. Others will not. And not all incidences of the presence of the short allele will evoke depression. There is no strict law. Is there a causal explanation for why some adults suffer from depression and others do not? The answer must surely be yes. Although this case of complexity does not fit the standard view of causal explanations that require universal laws (Hempel and Oppenheim 1948), causal knowledge of how and when genes contribute to the disease offer explanatory and predictive insight. To understand how that works, we need to move beyond the confines of the strictest notion of lawfulness and engage with the different degrees of stability, robustness, and contingency that the systems studied by modern science detect.

4

SCIENCE

HOW WE INVESTIGATE THE WORLD

I argued in chapter 2 that complexity of composition occasions the possibility of emergence, with its dynamically complex processes, and requires an expansion of our conceptual framework to include properties that are not amenable to simple reductive understanding. Evolved complexity, as I discussed in chapter 3, challenges the traditional contours of our concept of scientific law and encourages an expansion of lawful knowledge to include explanations by causal arrangements fraught with exceptions and contingency. In this chapter I will explore the methodological consequences of complexity for experimentation and causal inference. What kinds of complexity are revealed by new techniques of investigating complex structures? What assumptions about causal inference and causation itself are brought into sharp relief by these developments? Complex structures and their behavior are contingent on more or less stable setups scattered around the universe. Context matters. I will argue that accepting contingency and context-sensitive causal behavior as part of the fabric of the world threatens a simple and monistic notion of "the scientific method." Pragmatic and pluralist approaches to a multiplicity of scientific methodologies provide better scaffolding for an integrated understanding of our complex world.

Knockout Experiments

Here I will explore scientific approaches to understanding the causal relationship between genes and the properties of the organism they influence, that is, the organism's phenotypic traits like eye color or susceptibility to episodes of adult depression. It should be no surprise that the relationship is complex.[1] But as should be clear by now, that fact alone doesn't necessarily rule out successful reductive causal explanations; it all depends on the nature of the complexity. For decades researchers tried to create tools that allowed for precise control over a specific gene in order to study its function in controlling phenotypic expression. A breakthrough technology was developed in the 1980s known as transgenics or gene transfer (Müller 1999; Wolff and Lederberg 1994; Nelson 1997; Wolfer 2002). The technique involves intentionally introducing a foreign replacement gene into an organism in an attempt to reveal the function of the gene it replaces. In transgenic experiments, the objective is disruption of effects produced by the original gene. The hoped-for outcome is information about the phenotypic effects of the disrupted gene. This kind of investigation of cause and effect appears straightforward: observe the situation with and without the alleged cause. It is just like observing the behavior of a charged particle with and without the influence of a magnetic field. The difference in the observed effects should tell us something about the power of the cause we vary.

Early transgenics could not control where in the genome the new genetic material would attach, and since position is important to function, they were not particularly good indicators of the causal relationship between a particular gene and a specific phenotype. Further innovations developed, known as genetic knockout techniques. It became possible to aim the inserted gene at a precise location in, for example, the mouse genome. This gave scientists the ability to replace, or knock out, a specific gene with an inactive mutated sequence of DNA (Müller 1999). The gene knockout is created by selectively disabling a specific target gene in the embryonic stem cells. If the embryo with the knocked-out gene is viable and matures, it can be bred with another mouse subjected to an identical gene knockout protocol to create double-mutant offspring. These animals have two inactive genes in the location of the two normal alleles. Thus knockout animals provide an investigative technique that allows a particular gene of interest to be effectively removed. The morphology and physiology of the normal animal and the double-mutant animal are then compared to determine exactly what effect the normal gene has in the life of an organism.

Knockout experiments approach the ideal type of controlled experiment that is taken to grant the strongest inferences concerning causal structure. Such inferences are modeled in the famous logical rules that John Stuart Mill, following John Herschel following Francis Bacon, codified in the so-called Mill's methods (Mill 1843). Mill's method of difference, compared with the other methods, provides the clearest causal inference. If you can observe two systems that differ with respect to only one factor, all others being in agreement, then the differences in the effects between those two systems can be attributed to the differing test factor. Most experimental design is built on the edifice of Mill's methods of inference. Knockout experiments make use of this form of inference.

What do we learn from knockout experiments? In about 15% of the cases, the intervention on the knocked-out gene turns out to be lethal. Thus one knows that the gene plays an important role in the development and survival of the organism, but its precise function cannot be identified. In roughly half the experiments, specific phenotypic differences are observed in the offspring of the normal and the mutant, and the targeted gene's function is inferred. Much of this research investigates genes involved in serious medical conditions. The p53 knockout mouse, for example, has shown that when the *TP53* gene, which codes for a protein that is important in cell growth, is knocked out in a mouse, the animal develops numerous tumors. Humans who have a mutation at the *TP53* gene suffer from Li-Fraumeni syndrome, a rare, dominant genetic condition that greatly increases the risk of developing bone cancers, breast cancer, and blood cancers at an early age (Li and Fraumeni 1969). The presence of the normal *TP53* gene contributes to normal cell development, as long as no other source of cell damage is active. The absence of the *TP53* gene in the form of a mutation leads to abnormal cell development attended by high susceptibility for cancerous tumors. In this case the knockout technique did what it is supposed to; expose a general causal effect of the *TP53* gene.[2]

However, in about 30% of knockout experiments with viable double mutants, there is little or no evident phenotypic consequence of knocking out a gene (Ihle 2000; Edelman and Gally 2001). What is going on? If the gene causes a trait and you block the gene, shouldn't you affect some trait? Is the knocked-out gene not causally responsible for any trait? The logic of controlled experiments suggests that if the gene that is knocked out is the cause of a trait and you successfully block the causal contribution of that gene, then the trait product must be affected. One plausible conclusion is that in the 30% of knockout experiments with no phenotypic effect, the blocked gene is not causally relevant to producing the targeted trait. But the results are not always clear-cut. Some geneticists claim the cases where the knockout produces no substantive phenotypic difference actually point to the existence of a more complex causal network and in particular the dynamic plasticity of the genetic network (Edelman and Gally 2001; Greenspan 2001).

The confusion regarding what to make of anomalous experimental results when double knockout and normal organisms display very similar phenotypes is evident in the biological community. Two passages from the literature indicate some of the different perspectives. The first is by one of the inventors of the knockout technique who received the Nobel Prize in 2007 for his research in this area. Mario Capecchi claims that there *must* be a difference if a gene is knocked out. "I don't believe in complete redundancy. If we knock out a gene and don't see something, we're not looking correctly" (quoted in Travis 1992, 1394). The other, by Robert Weinberg, a pioneer in the study of the genetics of cancer, suggests that if there is no difference between the normal and the knockout, the inference should be that there is no significant function for the individual knocked-out gene. "The big surprise to date is that so many individual genes, each of which has been thought important, have been found to be nonessential for development" (quoted in Travis 1992, 1394). What is responsible for this confusion?

The results of knockout gene techniques are difficult to interpret because most effects are sensitive to genetic background, locally and globally. In traits whose genes are recessive, for example, sickle-cell anemia, the development of the trait depends on having *two* copies of the gene at a given locus. One copy in the company of a different allele will not generate the disease; indeed the heterozygous state produces resistance to malaria, a situation that helps explain why the sickle-cell gene persists in human populations (Pauling, Singer, and Wells 1949). In the case of dominance, as in Li-Fraumeni syndrome, a single copy of the gene causes the effect whatever the value of the other pair at the locus. Genetic causation is context sensitive.

What is referenced by the term *gene* has become more complex. The concept of the gene has evolved from its first use in 1901 by Johanssen to characterize what Mendel had called a genetic factor. At first, the gene was identified functionally, as whatever it was internal to the organism that was inherited and responsible for the differences in phenotypic traits, like flower color being either white or red. With the development of microscopic techniques, the gene was localized on the chromosome within the nucleus of a cell, and in 1953 Watson and Crick famously discovered the double-helix structure of DNA that constitutes the gene. Given the new techniques and information about genetic structure and function, in contemporary biology and philosophy there is much discussion about exactly what is meant by a gene (see Falk 1996; Griffiths and Stotz 2007). The classical view of a gene as a unit of inheritance located on a chromosome and coding for one protein is being replaced by more complicated compilations of DNA that include coding regions, repressors, and activator regulators (see Gerstein et al. 2007).

Single, simple sequences of DNA can issue in a trait, as discussed above, but in more complex contexts, the presence or absence of DNA will sometimes cause the effect in question and sometimes not, depending on what other nonstructural genes are present. This is the case for the well-known example of *lac* operon in *E. coli* studied by Jacques Monod and François Jacob (1961). In this case, the relevant DNA in the genetic system is comprised of operator, repressor, and promoter regions in addition to three structural genes. Lactose is metabolized when the three structural genes are expressed, but their expression depends on whether a repressor is bound to the regulatory site on the genome. Just having the structural genes, rather than some mutant forms, is insufficient to determine the behavior of the system. In addition, one needs to know the organization of the components and the other genes present in the context.

There are often multiple stages in the causal process from gene to protein to cell to other more distant phenotypic expressions that may make it difficult to ascribe a role to the presence or absence of an inherited gene.[3] What happens later on in the development and behavior of the organism is going to invite in yet more context-specific contributions along the way, so that the exact upshot of a knocked-out gene may be hard to pinpoint. Pleiotropy, when one gene has multiple effects, and epistasis, when multiple genes interact to produce an individual phenotypic effect, are well-known features of genetic complexity. But the "failed" knockout experiments arguably point to yet another type of complexity. These experiments indicate not just the existence of multiple causal components, but the complex *organization* of multiple causal components. Complex organization can display either redundancy (also referred to as copy redundancy) or robustness (also referred to as degeneracy). These types of architecture can produce significant problems for interpreting knockout experiments and inferring causal role.

Redundancy and Robustness

Redundancy occurs when a gene is knocked out but copies of the same gene are activated. This is equivalent to the engineering notion of a built-in "fail-safe" mechanism. If in a knockout experiment only one of multiple, redundant copies of the genes that code for a certain protein are knocked out, then the phenotype will be preserved (Wagner 2005, chapter 15). Although copy redundancy does occur in biological systems, it is somewhat puzzling how it could have evolved. Presumably, to carry around an extra copy of a functional component incurs a cost. There will be no correlative adaptive benefit unless the copy is at least occasionally called upon to operate. Nevertheless, it has been argued that there are scenarios in which copy redundancy could evolve (Nowak et al. 1997).

Robustness in a system, or "degeneracy" as defined by Edelman and Gally (2001), describes a different organizational complexity. If the genetic system is robust, then when a gene is knocked out, *other elements* of the genetic structure respond flexibly to generate a functional outcome similar to the normal system. Alternative components with distinct functions—for example, different genes in a genetic regulatory network, or different proteins in a cell signaling cascade—produce the effect of the targeted component when that component is no longer operative. Andreas Wagner (2005) refers to this as distributed robustness and provides a number of examples to argue for its prevalence over copy redundancy in biological systems. Wagner also presents cogent arguments for how and why robustness against mutation in complex biological systems would evolve, although he does admit that "there may be no fundamental theory of how robustness evolves, if such a theory is required to take into account the different architectures of biological systems" (Wagner 2005, 268; see also Ciliberti, Martin, and Wagner 2007). Robustness is not rare in biological systems. It has been identified at twenty-two different structural levels, including protein folding, biosynthetic and catalytic pathways, immune response, neural circuitry, and sensory modalities (Edelman and Gally 2001). In the words of one biologist, "Robustness is a property that allows a system to maintain its functions despite external and internal perturbations. It is one of the fundamental and ubiquitously observed systems-level phenomena that cannot be understood by looking at the individual components. A system must be robust to function in unpredictable environments using unreliable components" (Kitano 2004, 826).

The absence of a phenotypic change even when all the redundant copies of a single genetic component are knocked out could indicate that the network itself has reorganized to compensate for the loss of the gene. If so, parts of the network that in the normal state would be described by one set of functional relationships *change* their interactions in response to the experimental intervention to produce a product similar to that of the unperturbed system. Nodes that are connected in the normal network have different connections in the mutant knockout network, with both organizations generating the same output. Ralph Greenspan offers a model of what that could look like—see figure 4. He has also conducted experiments to document the applicability of the model of reorganizing genetic networks (2001; van Swinderen and Greenspan 2005). The genetic network affecting coordination in fruit-fly behavior was investigated. The pattern of epistatic interactions of sixteen genes was shown to vary dramatically in two different genetic contexts.

The capacity for robustness appears to be both ubiquitous and significant for biological systems. Indeed robustness gives evolved complex

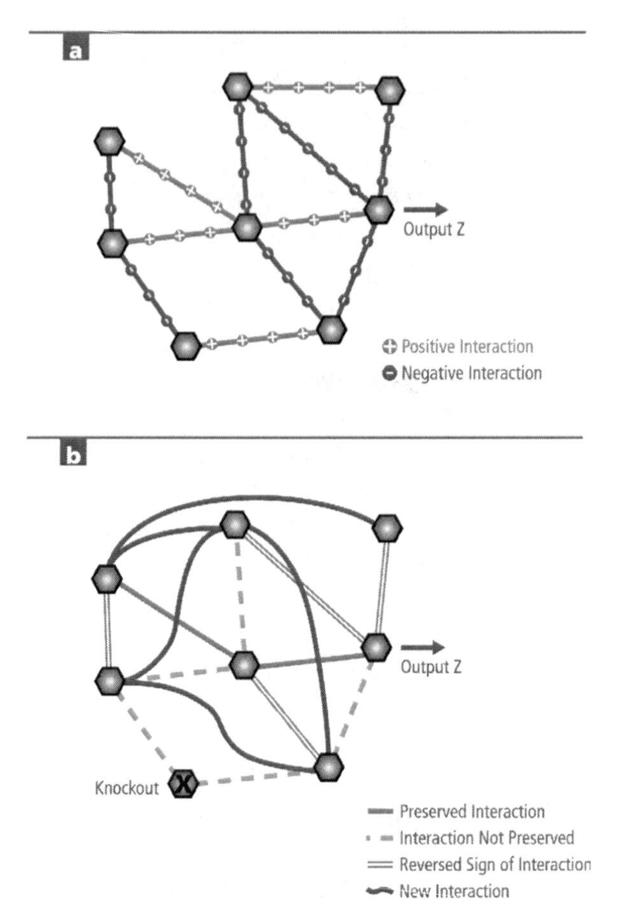

Figure 4. Flexible genome

systems, including organisms and ecosystems, the ability to survive in an environment with changing internal and external conditions. Let's look at a well-known example. In 1999 Alon and his collaborators described the robustness exhibited by the mechanisms of a widely studied model organism, the bacteria *E. coli*, as it copes with changes in the external environment. *E. coli* swims toward an attractant chemical and away from repellent chemicals in the environment. This is called chemotaxis. It is achieved through protein signaling networks where chemical receptors on the surface of the bacterium interact with elements within the cell to change the rate of a tumbling motion that moves the organism off a direct course. When the *E. coli* meets an attractive chemical, it stays on course, rather than tumbling into other regions that might not be so desirable. If the environment changes to a repellent chemical, the bacteria increases its tumbling to get away. Research has shown that this ability of the bacteria to track its chemical environment is robust to changes in the intracellular proteins that are key contributors to the network that inputs environmental chemicals and outputs the motor behavior of the flagella that makes the bacterium swim or tumble. Rather than a fine-tuned machine where every part has a precise role and all must be in order for it to operate, the network still delivers the correct tracking behavior when key proteins are varied either in concentration or by mutation. Thus the *E. coli*'s ability to match its behavior to the ambient environment by increasing or decreasing its tumbling behavior is managed by a robust internal biochemical network (see also Barkai and Leibler 1997; Stelling et al. 2004). Disrupting at least some of the internal components that play a role in the normal generation of the behavior does not annihilate the adaptive response to environmental stimuli.

Robustness characterizes mechanisms that produce flexible behavior as in chemotaxis as well as mechanisms that keep some property steady. The latter is found in homeostasis, for example, in the maintenance of stable body temperature in humans by sweating or shivering. That complex biological systems are robust to changes of external and internal conditions has been known for a long time. How they achieve this system-level property and how we can experiment to determine the mechanisms for robustness challenge simple concepts of cause and effect.

The concern raised is with the kind of stability that is required of causes. Is the simple stability of "same cause, same effect" implied by Mill's methods sufficient to capture this type of complex behavior? In particular, can the behavior of a system with multiple components organized in a robust network always be reduced to a sequence of simple causes, each separately contributing to the overall complex effect? The confusion of inferences from some knockout experiments generates a need for representing and analyzing the peculiar complexity of context-rich, dynamically responsive causal networks. It is not always easy or possible to isolate linear causal paths by shielding target causes from the effects of context. The traditional approach of decomposing complex systems into simple, modular component causes and then analytically reassembling them does not work straightforwardly in cases of robustness.[4] This presents a domain of science in which Mill's methods cannot be used in an algorithmic way because the causal role of a component depends in nonsimple ways on the context.

Interventionist Causes and Modularity

The types of complexity found in gene-trait relationships that contribute to problems for drawing inferences from knockout experiments have implications that go beyond molecular biology. Indeed, given the naturalistic approach to philosophy adopted in this book, the impact may be felt by our very understanding of the concept of cause. As Searle aptly put it (1999, 2069), "The question 'What is the cause of cancer?' is a scientific and not a philosophical question. The question 'What is the nature of causation?' is a philosophical and not a scientific question." Yet these two questions are related. If contemporary scientific explanations of complex biological causation fail to be accounted for by our philosophical theories of cause or causal inference, then something has to change. Either the scientific use has to be brought to conform to philosophical analysis, or the philosophical theory of cause has to expand to include what is captured by the scientific use. In this section, I will consider what I take to be the most promising theory of causation in contemporary philosophy of science, and the consequences for it when confronted by complex biological causal behavior.

An interventionist account of causal explanation, recently developed by James Woodward (2003), is a sophisticated intellectual heir of Mill's methods that focuses on how disruption of a cause, or intervention on a variable, makes a difference. Woodward's account is both a reflection on the methods used to detect causes, and a theory of what we mean by "cause." If changes in one factor induce corresponding changes in another factor, the first is a candidate for being the cause of the second. What is required in order for causes to be explanatory is that the cause-effect relationship is invariant or stable, but the relationship need not hold universally (see Woodward 2001 and Mitchell 2002a for a discussion of the relationship of these two notions of invariance and stability). That is, invariance of the causal relationship need only hold in various, but not necessarily all, contexts. This view fits well with the account of laws I defended in chapter 3. Invariance describes a *property* of the relationship between the cause and effect or the independent and dependent variable of a function $F(X, Y)$. A strong example of such a function is Newton's law of gravity, $F = G(M_1M_2/R^2)$. The force is determined by the masses and distance between two bodies. A function is invariant if under certain "ideal interventions" where the value of X changes, the function accurately describes the resulting value of Y. Basically, when you "wiggle" X and nothing else changes, then Y "wiggles" in a regular, functionally describable, determinate manner. In the case of Newton's law, change the distance, and the force changes in the way specified by the function. If that is the case, on Woodward's account, the changes in X cause the changes in Y, and the function $F(X, Y)$ explains why and how Y changes. Of course, Newton's law does break down in some contexts. On Woodward's account, this doesn't threaten the validity of the causal description.

For X to be a cause and the causal functional relationship to be explanatory, it is not necessary for the relationship to be universal and exceptionless. It need happen only in a range of values for X. Giving up universality permits us to see that invariance comes in degrees. Woodward allows that some causal relations are more invariant, others less. Consider Maxwell's equations, which in fact explain the existence and properties of electromagnetic radiation. On Maxwell's theory light is emitted by accelerated charges; hence it should be emitted by the electrons in orbit around the nucleus of an atom. A well-known challenge for those developing atomic theory at the beginning of the twentieth century is the fact that Maxwell's equations fail in the case of an orbiting electron as depicted in Bohr's model of an atom. On Maxwell's theory the electrons in the Bohr model of the atom, which are accelerating by virtue of the fact that they are constantly changing direction as they orbit the nucleus, would emit light, continuously losing energy. They would quickly collapse into the nucleus and make the atoms unstable. This doesn't happen, and this fact led Bohr to postulate the existence of "stationary states" for these electrons that are exceptions to Maxwell's equations. Thus Maxwell's theory is widely invariant but not universal; that is, it does not describe the behavior of electrons in Bohr's model of the atom. As was described in chapter 3, laws display differing degrees of invariance.

Consider the equations describing the relationship between volume, temperature, and pressure of an ideal gas, $PV = nRT$ (where P is the pressure, V the volume, and T the temperature. R is a universal constant, and n is the number of moles). The behavior of real gases usually agrees with the predictions of the ideal gas equation to within 5% at normal temperatures and pressures. At low temperatures or high pressures, real gases deviate significantly. Ohm's law, a regularity used daily by electrical engineers that relates voltage and current, has long been known to fail in semiconducting materials and in situations involving strong electric fields. Biological laws, such as Mendel's law of segregation of genetic factors during gamete formation, are certainly even less invariant, that is, there are more circumstances in which they will fail. As a conception of causation, invariance under intervention, or stability upon some set of conditions, is closer to the type of causal relationships found in biology than the universal, exceptionless requirements of traditional epistemologies.

There are, however, aspects of Woodward's account of causal explanation that become more difficult to reconcile with contemporary investigations of complexity. The problems revolve in part around his notion of "modularity," which he identifies as a second type of stability that true causes possess. Modularity applies to a collection of causal functions, rather than the variables in a specific function, and occurs in causal systems where multiple causal functions are simultaneously operating. Modularity denotes the property of separability of the different causal contributions to an overall effect. In simple cases the modularity condition is usually met. If two Newtonian forces act on a single body, say gravitation and friction, then the effects of their actions are separable. One can attribute some aspect of the final motion as due to friction, the other to gravity. If there were no friction, the effect due to gravity would be unchanged, and vice versa. To get the overall effect in this case, we use the vector sum of the forces to predict the motion that will result from their simultaneous operation. In his original papers on causality, Woodward argued that modularity, like invariance, is essential to a functional relationship being a causal relationship: "This is implicit in the way people think about causation . . . this sort of independence is *essential* to the notion of causation. Causation is connected to manipulability and that connection entails that *separate mechanisms are in principle independently disruptable*" (Hausman and Woodward 1999, 550; my emphasis).

Modularity entails that true causes, even in complex situations, can be separated out and the total effect can be expressed as some sort of well-defined composition of the separate causes. In addition, for Woodward, modularity of causes means that we can intervene to study a single causal relationship in a complex situation without that intervention affecting the other causes that are operating simultaneously. They are, in Woodward's terms, separately disruptable. Modularity therefore not only defines the nature of true causes but also suggests how scientists can study causal relations: each causal factor can be studied in isolation.

Since the behavior of complex biological systems depends on multiple causes, it is reasonable to question their separability. Are all causes in biological systems "modular" in Woodward's sense? The worry is about how strong to take the requirement of modularity in order to ascribe causality to a component of a complex causal network (for discussions of problems with the modularity requirement, see also Cartwright 2002; Bogen 2004; Mitchell 2008c).

To examine this issue, let's return to the anomalous knockout experiments. How do they fare on Woodward's conception of casual explanation? It appears that a degenerate or robust system where a genetic network reorganizes when some piece of it is knocked out is not independently disruptable. That is, one gene in the network functions as a casual contribution to the phenotypic effect under normal internal conditions, but functions in a different way, when another part of the network is removed (see figure 4). Thus Woodward's condition of modularity is not met. Yet would we want to conclude that because the normal genetic structure could not be independently disrupted that the knocked-out gene was not a cause contributing to the phenotype in the normal organism? Probably not. While some causally invariant relationships are modular in this strong sense, it appears that others are not, yet I believe we would still want to identify them as both causal and explanatory. To quote Greenspan (2001, 384),

> The . . . proliferation of identified genes creates problems of interpretation; such as which genes are most important and how they all interconnect. . . . Not the least of these problems is the diminishing value of the pathway analogy. Rather than running in linear paths, the increasing complexity of relationships among genes is better described as a distributed network. Some genes produce more damage than others when mutated, but this depends heavily on the context of other alleles that are present, and so it is difficult to arrange them in a simple order of importance. The interactions that have defined various pathways are not wrong, just not exhaustive. They are part of a much larger picture.

How do we resolve the problem raised by a requirement of modularity or independent separability of causes, and the fact of reorganizing, nonmodular complex networks? I believe there are three possible responses.

1. We can infer that the genes in the normal genetic pathways are *not* causes because they fail to satisfy the condition of modularity. This is a "bite the bullet" strategy where if a scientific discovery does not meet the criteria of our philosophical viewpoint, we hold fast to the view and deny the scientific discovery. This is not a solution I can recommend, given my naturalistic strategy.
2. We can redescribe the network in finer or coarser granularity to try to satisfy the modularity/independence condition. This is a "make the world fit your theory" strategy, which is a version of "bite the bullet" using representational redescription to try to save the philosophical view. It at least takes seriously the scientific discovery and attempts to mold the results in some way to preserve both. This strategy has some attraction, but the results, as I shall argue, will not give us a uniform representational grain size for analysis to use across all behaviors of a system type.
3. We can infer that modular causes do not exhaust all the types of causality found in nature. This is the pragmatic, pluralist, integrative strategy, and it is the one I will defend.

Why is the "bite the bullet" strategy undesirable? To say that the knocked-out gene is not a cause in the normal case just because in some knockout cases other causes step in to ensure the robustness of the system in the face of internal perturbation ignores the fact that there can be, and often are, multiple pathways that generate the same outcome. Many systems display this type of causal behavior.

Let's consider the case of brain reorganization to see why a strong adherence to modularity is puzzling. Since the 1980s a number of different experiments have been done in adult mammals to investigate the consequences of injury to parts of the brain or nerves that in the normal adult perform standard functions, for example, in language skills or sensory perception. After the removal or destruction of brain tissue or entire regions of nerve input, it is shown that the brain reorganizes to assign new regions in the brain to produce the same functional outcome of the removed part. Ramachandran's work on phantom limbs (1993) is illustrative.

How could an amputee feel pain or any other sensation in a limb that is no longer there? Ramachandran proposed that the areas in the brain near those that had been and are typically stimulated by the nerves in an existing hand yet no longer receive stimulation due to the amputation are remapped to express the referred pain. These are the sensory areas of the face and the arm. Thus when patients are subject to, for example, a trickle of water down the face, they feel it as a trickle of water down their phantom hand, and specific points on the face referred to specific digits of the amputated hand.

> Because the hand region of the cortex was now deprived of input (due to the amputation), it was "hungry" for new input, so the sensory input normally destined exclusively for the adjacent hand cortex now "invades" the vacated territory. Thus, a touch of the face is misinterpreted by higher brain centers as arising from the hand. . . . [T]his was the first demonstration of such large-scale reorganization of sensory maps in the adult human cortex. (Ramachandran 2007, 406)

This example, I believe, is similar to the genetic knockout case where once a part of a normal network is removed—either the gene in the regulatory gene-phenotype causal network or the sensing nerves of the hand in the limb-brain sensation network—other structures that had other functions take over the causal role of the missing ones to issue in similar effects by means of different pathways. I find it hard to accept an account that holds that, because sensations of the face are now experienced in the area of the brain normally associated with the hand, in the normal case the area of the brain that typically is involved in the causal structure of sensitivity is not truly involved in the cause of pain. When reorganization occurs, that does not negate the causal role of the missing structure in the normal case, even though by reorganizing the causes making up the complex structure it fails to be modular, or independently separable.

Maybe the "make the world fit the theory" strategy will fare better. Perhaps we have not correctly described the causal structure to capture all of its causal relations. A complete description might include not only the causes active in the normal operation of a genetic network or of the nerves and brain network but also those that are possible given any internal or external perturbation to the system. If that were the case, then the wiring diagram would be complicated, but the parts of it might be separable.

This solution misconstrues the character of explanation that involves complexities of the kind we find in biology. As I have suggested above, most explanatory projects in biology are directed not at what might be possible given the building blocks of physical and chemical components, but rather at what has actually evolved, given the role of phylogenetic history, developmental constraint, adaptation, and chance. There is a range of possibilities that is explored and explained, but it is generally not the entire space of what is biologically possible. Given the contingency of the forms and behaviors of biological objects, describing all the physically and chemically possible interactions does not zero in on the actual target of biological explanation. What regions of possibility space must be represented to give the complete causal map? If the answer is "all of it," then this strategy would be cognitively intractable. While the actual physical and chemical structures constrain what could have evolved, they do not determine what has in fact evolved. Rather, that is the result of contingent evolutionary history, replete with undirected randomness.

Biologists seek explanations for the actual domain of biological forms and behaviors, and not the set of causal functions that are true of the larger set of the biologically possible forms. The target of explanation is the particular process by which some effect has, in fact, been brought about. There are often many pathways by which a specified effect could come about, and robustness ensures that if perturbed from its normal state, the biological system can recover something close to its normal function by finding a new pathway. Since what a component of a complex genetic network does is dependent on the context, it may be too sensitive to be reliably discovered by means of simple one-shot perturbation studies.

Perhaps we do not need to consider and represent all possible causes to restore modularity to our explanatory causes. Perhaps the system itself is a modular cause, and not the separate genetic pathways that constitute it. Thus modularity in Woodward's sense could be taken to impose a constraint on identifying a cause. However, this strategy also has some odd consequences. By limiting our notion of cause to "well-behaved" functions that are modular, we will on some occasions identify a cause at one level (e.g. the gene) while in other circumstances that incorporate the same component, identify the cause at a different level (the genetic network, or the genes plus environment nexus). A stretch of DNA might be a cause in some contexts, but not in others.

In sum, modularity is the mark of a type of independence from context. The same functional relationship between variables will hold for a given component of the contributing mechanisms whether or not there is a change in a different component. The total effect may change when different components contribute, but the operation of the modular mechanism will not be changed nor change its causal contribution. In situations where the presence or absence of other contributing factors changes the behavior of a component, and not just the total effect, modularity will fail. In the knockout experiments with no difference between normal organisms and knockouts under robust reorganization, modularity fails for individual genetic components, and hence they cannot be identified as causes of the normal phenotype. But what about the knockout experiments that are successful? In these cases the single perturbation study does reveal a significant causal contribution of a single gene. Thus if we adopt the second strategy, that is, assign the cause to the grain size that satisfies the modularity condition, there will not be just one correct level of organization in which to analyze causal invariance and from which to generalize or export to other cases. When single knockout experiments work, when the system is well behaved, then it does display invariance and modularity, and identifying causes may be relatively straightforward. The "gene for" phenylketonuria, or PKU disease, approximates this case where a naturally occurring double mutant at a specific site on chromosome 12 in humans prevents the production of the enzyme phenylalanine hydroxylase (see Ottman and Rao 1990). In the normal state the enzyme is produced; in the mutated state it is not. Here the gene is the right level of granularity to attribute causality.

In robust, reorganizing networks, knocking out a single gene does not always issue in a change in the phenotypic effect. Thus genes are sometimes the right level to find modularity, but not always. Wagner (1999) has argued that if the behavior of a complex system is linear, then it will always be possible to find a representation such that the effects of the components become independent of the context of the other variables. However, if the system is nonlinear, which is often the case in biology, then one "may need to know the state of the entire system with all its state variables to make predictions" (Wagner 1999, 99). Any physical system with complex feedback mechanisms will be one in which we can expect modularity to fail. But we should not conclude that such systems don't involve true component causes.

I have argued that some forms of complex organization and dynamics present problems even for Woodward's more forgiving approach for causal explanation (Mitchell 2008c). In particular, when a system is compositionally integrative in the Bechtel and Richardson sense (1993), in which the causal properties of the parts are themselves nonlinearly dependent on the whole network, then modularity or independent disruptability will not be satisfied. If we read the modularity requirement strongly, as defining what it is for a component to make a causal contribution to the system's behavior, then we are left with the uncomfortable conclusion that a component may have a causal role in an unperturbed system, but loses its causal status when the perturbed system reorganizes. One way out of this situation is to let the individuation of causes be constrained by modularity and accept that the appropriate level of organization to find causal explanations is itself context dependent.

The pragmatic alternative is to take modularity to be a feature present in some but not all causal systems. Clearly with modularity the discovery of causal structure becomes easier, for controlled experiments eliminating single components will always allow their causal contribution to be discerned. Individual causal chains that make up a more complex causal network can be studied in isolation from the rest, and they will make the same contribution no matter what other causal factors enter into one or another system. But not all causal structures are that well behaved. Some complex structures harbor nonmodular, context-sensitive actual causes that can explain their behavior. The methodological consequences of failures of modularity are significant. Scientists have to engage in analysis of nonmodular feedback networks to push science forward. This is being done by multivariate experimental techniques and network analysis. Exploring multiple scenarios rather than shielded experiments on targeted single components may prove more conducive to understanding context-rich casual structures. Series of pairwise knockout experiments can expose which nodes in the causal network interact in which ways and where these interactions are expressed in the development of the organism (Solloway and Robertson 1999). In addition, a "multifactorial perturbation" approach, "in which gene variants and environmental factors are present in many combinations" (Jansen 2003), may be able to disentangle the structure of redundant and robust networks. The problems of discovering the structure of interacting genes are serious. With humans having roughly thirty thousand genes, assuming each gene has only two states, there are $2^{30,000}$ multigene states. Testing all combinations of states of the genome to discover which combinations affect the phenotype clearly is not feasible. There are ways to manage the complexity, including using statistical data from microarray analysis to develop structural equation models.[5] The techniques for doing so are tested against simulations where the actual structures are known, and if they perform well there, then one has reason to trust the network models that are generated from experimental data.

A new field of "systems biology" that integrates simulation studies, experimental results, and statistical analysis is evolving to meet the demands of describing complex networks (Kitano 2002; Sauer, Heinemann, and Zamboni 2007; Boogerd et al. 2007). This new field embraces the study of the whole system, incorporating an approach that is nonreductive and does not assume that modularity holds, targeting emergent properties that could not be predicted by studying the components in isolation. Multiple variables are investigated simultaneously in varying contexts, permitting a view into an entire system with an aim to disentangle the dynamic structure that, in the end, is causing the properties we are interested in explaining. "The reductionist approach has successfully identified most of the components and many interactions but, unfortunately, offers no convincing concepts and methods to comprehend how system properties emerge" (Sauer, Heinemann, and Zamboni 2007, 550). A recent study by Ishii et al. (2007) on *E. coli* serves as an example of the new approach to understanding biological complexity. They studied 24 mutant strains of *E. coli* in which a different gene that functions in carbon metabolism was removed from each strain. They looked at three levels of organization in the bacteria: gene, protein, and metabolites (the products of the carbon metabolism system). They discovered that metabolic rate for growth changed in light of environmental changes, but was robust to changes at the genetic level. In contrast, single-gene knockout experiments would show no change in metabolite products, leaving the question of its causal role undetermined. Multilevel, multicomponent integrated approaches to complex structures offers new methods for understanding the world we live in.

Philosophy needs to adjust the concepts of what counts as a cause and the logic of causal inference to match the developments in the scientific study of complex natural systems. The feedback loops and multiple kinds of contingencies that structure complex systems were not taken into account in Mill's methods. There is no single way to acquire knowledge of the causal structure of our world, no single "scientific method," since there are multiple types of causation exhibited by the evolved, contingent systems that make up nature. There must be (as there in fact are) a plurality of scientific methodologies that embrace richer possibilities about the causal situations being studied.

5

POLICY

HOW WE ACT IN THE WORLD

Perhaps the most important way in which a better understanding of complexity may revise our thinking about the world is in making decisions and forging policies. In both individual choice and social policy contexts, we consider the consequences of our proposed actions along with the values attached to those consequences in order to determine which actions we expect will best promote our values. For decades political scientists, economists, logicians, decision theorists, and others have engaged in the analysis and modeling of human decision making, presenting formal models using probability theory and automating those models in computer programs (Savage 1954; Simon 1960; von Wright 1963; Rapoport 1989; Pratt, Raiffa, and Schlaifer 1995). Indeed, their models and associated decision-making methodologies are often employed by politicians and regulators in efforts to make better decisions. Having a simple and effective calculus for deciding on the best actions would be a great boon for humanity. Unfortunately, the uncertainties in a world of context-sensitive, dynamically responsive complexity present substantial challenges to aspects of the standard methods that scholars have developed for modeling and informing decision making. New ways of thinking about decisions and effective policy configuration may be required if we are to choose wisely.

The standard strategy for policymaking has been predict-and-act models. In these models all decisions take into account two fundamental components: the projected outcomes of our actions, and the values we assign to those outcomes. Analyses of these components issue in a choice among alternative actions. In many circumstances, both the factual and the value inputs to decision making involve complexities that strain the capacity of current models to guide us adequately. In textbook cases the simplest circumstances—that is, decisions under certainty—are presented to clarify the logic of decision making, and then the reasoning is extended to contexts of risk and uncertainty (Jeffrey 1990). If each possible action has a single certain outcome—say, growing corn will yield a profit of X and growing soybeans will yield a profit of Y, and X > Y—then, given one generally prefers to make more money in an economic enterprise, one should rationally choose the action of growing corn. Of course, even decisions with certain outcomes can have complications due to weak preference structures or branching downstream consequences with multiple decision nodes. But things can be much more complicated. In many decision contexts, certainty about factual outcomes is not even a distant dream. This is the case *especially* with respect to choices that involve future states influenced by complex processes in nature and in society.

Decisions under conditions of risk acknowledge the absence of certainty, and rely instead on probabilities attached to outcomes. The models developed to help us make better decisions in these contexts apply various forms of maximization logic, that is, mathematical representations designed to help us take actions that will advance our values, even when the outcomes are not certain. In cases where an action may have different possible outcomes, for example, when the amount of crop yielded by growing corn depends on the weather, then the various outcomes (high yield, low yield) are assigned a probability. The same procedure is applied to outcomes associated with each alternative choice, in this case to growing soybeans. Instead of each distinct action that we are considering being associated with a single certain outcome with a single value, we consider all the outcomes of each choice and their associated values consolidated in a measure of expected value. The expected benefit (or in the language of probability theory the "expectation value") of an action is the sum of the possible outcome values weighted by their probabilities. If soybeans are more productive than corn when weather is bad and the probability of the weather being bad is high, then the expected value of growing soybeans may well be higher than the expected value for growing corn, even though one would actually do better growing corn in the less likely situation of the weather being good. Ideas about using expectation values for decision making have been around since Bernoulli posed the St. Petersburg paradox in 1713 (Krüger, Daston, and Heidelberger 1990).[1]

The subjects of decision theory, risk analysis, and policymaking constitute a vast and developing area of scholarly and practical research. There are important critiques of the simple traditional model described above, from both normative and descriptive perspectives (Kahneman and Tversky 1973; Gigerenzer 1996; Vranas 2000). New formal representations that take into account the complexities of serial and nonserial problem structure (Esogbue and Marks 1974), collective choice (Arrow 1951; Sandler 1993), implementing various interpretations of probability (Korb and Nicholson 2003; Peterson 2008), and new computational resources (Boutilier, Dean, and Hanks 1999) have been elaborated. What I offer here is not a comprehensive review of this literature and how the various theories cope with complexity. I have a more limited goal. I will present an argument about how one potential feature of uncertainty—namely, ineliminable risk—challenges the traditional framework for decisions under uncertainty. I will explore some means for meeting that challenge.

In the traditional view of decision under risk, the risks associated with each action are modeled by assigning quantitative probabilities to the range of outcomes that a given action may incur. Formal decision-making models are designed to tell us which actions will maximize benefit and minimize costs given the range of probable outcomes. This form of calculation to inform decision making is both the substance of introductory university classes in various social sciences and prescribed for policymakers by law. Cost-benefit analysis, taking into account the expectation values of various outcomes of actions we are considering, has become the standard for policymaking in the United States since 1981 when the Reagan administration introduced Executive Order 12291, mandating that "regulatory action shall not be undertaken unless the potential benefits to society for the regulation outweigh the potential costs to society; . . . and regulatory objectives shall be chosen to maximize the net benefits to society." In a recent publication of the Environmental Protection Agency, the *Ecological Benefits Assessment Strategic Plan* published in October 2006, the benefit-cost analysis (or BCA) is further clarified: "BCA estimates the net benefits to society as a whole by comparing the *expected* benefits accruing to those made better off by a policy to the *expected* costs imposed on those made worse off" (3; my emphasis). Using expectation values to represent the total outcome (of costs and benefits) to assess the desirability of policy options requires the assignment of quantitative probabilities to the various outcomes that are associated with each policy. The goal is to support a quite natural predict-and-act model for managing everything from large organizations to economies to ecosystems. The challenge is that the traditional modeling techniques depend on filling in a number of factors, including identifying possible future outcomes, assigning costs and benefits to them, and assigning quantitative probabilities to the projected outcomes. These tasks cannot be easily accomplished when we are facing situations involving complexity. Consequently, we need to rethink how we model uncertainty in these systems and rethink the predict-and-act model itself in designing policy strategies in a complex world. Achieving desired outcomes may require adopting more complex strategies for effective policies.

The goal of reducing policymaking to decisions computed by algorithms for maximizing benefit and minimizing cost faces a substantial barrier if we understand the complexity of the world in the ways that I have advocated in this book. Scientists acknowledge that the global weather system and the biosphere are both paradigms of a "complex system" in all the senses of complexity discussed in earlier sections. For such systems, assigning probabilities to, say, warming of average global temperatures by 10 degrees in 100 years if greenhouse gases are not reduced, or to the effects on wild species' extinction rates with the introduction of genetically modified crops, is virtually impossible. "Climate change presents a very real risk. It seems worth a very large premium to insure ourselves against the most catastrophic scenarios. Denying the risk seems utterly stupid. Claiming we can calculate the probabilities with any degree of skill seems equally stupid" (Prof. C. Wunsch, quoted in Revkin 2007). The reasons it is so difficult are the large number of variables that affect an outcome whose contributions must be both understood and measured, the role of variables we have not even identified that contribute to the outcome, and the intervention of randomness, that is, variables that are completely outside the system but that can affect the system behavior we are attempting to predict. Uncertainty about the probabilities of outcomes is pervasive, multiplicative, and often nonlinear in complex systems.

The predict-and-act model suggests that the strategy for improving the knowledge base for all decisions is to reduce the uncertainty to reach, if not certainty, then at least objective (or consensual) probability assignments. However, in cases of complex systems, it may very well be that waiting until there is agreement or confidence in the quantitative probability assigned to possible outcomes is unreasonable. For example, we may be waiting until it is too late to act to avoid seriously undesirable consequences. Indeed, depending on the structure of the system and the nature of the complexity, knowing more about a system may not in fact reduce the uncertainty. Walters (1997) argues that "we cannot assume that increasing model detail (more complete representation of space-time event structure) will result in progressively more accurate predictions and/or reduced risk of making a very bad prediction." One reason is that adding more detail typically adds more parameters to the model structure, and "each of these parameters is likely to be less well supported by field data; this 'overparameterization' can degrade the predictions of a mechanistic model" (Walters 1997, 1). Simply increasing our knowledge of the number of factors affecting the behavior of a complex system may not reduce uncertainty. Alternative representations of what is known and what is not known, and alternative policy strategies that acknowledge ineliminable uncertainty, promise to provide a better guide to decision making.

Facing up to complexity suggests a shift in the ways in which we represent decision situations. Sarewitz and Pielke (2000, 14) argue against what they call reductionist approaches for prediction in the earth sciences in favor of a pragmatic strategy that acknowledges inherent contingency in the systems studied: "Rather than identifying the invariant behavior of isolated natural phenomena, prediction of complex systems seeks to characterize the contingent relations among a large but finite number of such phenomena." They warn that relying on a reductionist notion of the ability of science to generate predictions may deter policymakers from using other, more appropriate guides for rational decision making. That is, "if decision makers lack data about present environmental trends, or lack insight into the implications of different policy scenarios, they are less likely to use adaptive approaches to environmental problems, and more likely to wait for a predictive 'prescription'" (20; see also Lacey 1999, 2002). But if the predict-and-act model cannot be implemented with the knowledge we have now, and if attempting to get it and waiting to make decisions once we have is not desirable, then how can we make reasoned decisions in contexts of deep uncertainty?

Scenario analysis and robustness comparison offer an alternative decision procedure to maximization of expected value based on quantitative probability assignments. The "predict" part of "predict and act" is replaced by a model of multiple alternative futures. Policies are judged by how well they perform over the entire scope of possible futures. Adaptive management offers an alternative policy architecture to the "act" part of "predict and act." It may be better able to cope with uncertain systems. Adaptive management, already employed in conservation efforts, takes into account the dynamics of the complex systems we engage with and are part of, as well as the changes in our knowledge of those systems. An all-or-nothing policy, a once-and-for-all-times decision, a "no surprises" clause to prevent modification of prior decisions, will fail to guide us in choosing rational actions that promote our values in a complex world (Shilling 1997; Wilhere 2002). I will now compare "predict and act" with "evaluate scenarios and adaptively manage."

How probable are the worst effects of increases in the carbon dioxide content of the atmosphere on global climate, or of introducing genetically modified foods in our farms and diets on the loss of biological diversity? There appears to be no scientific consensus on the precise quantitative assignments of likelihood of various good and bad effects that may accrue. The Intergovernmental Panel on Climate Change (IPCC) report published in 2007 estimates that if atmospheric concentrations of greenhouse gases double compared to pre-industrial levels, this would "likely" cause an average warming of around 3°C with a range of 2–4.5°C. The climateprediction.net project, which aims to test climate models in order to "improve methods to quantify uncertainties of climate projections and scenarios, including long-term ensemble simulations using complex models," estimates that the effects of doubling CO_2 would generate increases in surface temperature from below 2°C to more than 11°C. How likely are the extremes of this range of changes, or of any of the predictions in between? Even more important for decision making, how likely are any of these outcomes given the range of actions with respect to CO_2 production, from doing nothing to completely curtailing artificial sources of CO_2, that might be taken? The 2001 IPCC assigned a probability of 66% that observed increase in temperatures is due to the observed increase in greenhouse gas concentrations, while the 2007 IPCC report increased the probability of that claim to 90%. The fact is that the community of scientists, while reaching a consensus that human actions are responsible for global warming (Oreskes 2004), disagree about the precise quantitative probabilities associated with specific outcomes and, indeed, even with our knowledge of which processes underlie the assignments.

Scientific confidence in probabilities of various outcomes is changing as more data and better models are developed. It is not surprising that it is difficult to quantify the uncertainty into degrees of probability for models involving so many variables, whose impacts are sensitive to small variations in the values, and whose dynamics are rich with feedback (see Allen et al. 2000). Is the right strategy to focus exclusively on reducing uncertainty in the hopes of converging on a corroborated and clear numerical probability for the outcomes of actions before we act? The predict-and-act approach would suggest the answer is yes. In contrast, I argue that requiring agreement on a single quantitative assignment of probability to variables in complex systems actually runs the risk of undermining *any* scientific contribution to public policy decisions. This is a case where the best is the enemy of the good. Yearly (1996) has argued that policymakers use uncertainty and lack of scientific consensus to delay responses to environmental problems. Appeal to inevitable uncertainty creates an opportunity to challenge the credibility of any scientific claim, to the point of creating a policy environment in which all scientific contributions are dismissed.

Making policy decisions while ignoring scientific knowledge, even if that knowledge is deeply uncertain, is its own disaster. Single-probability assignments may well fail to represent what the scientific community does know, and for which there is substantial agreement, about uncertain future states of complex systems. Fortunately, there are alternative methods for representing uncertainty in complex systems, and they should change the way we think about policy strategies.

Robustness in Scenario Analysis

In the face of what may well be ineliminable uncertainty attaching to the predictions of the behavior of complex systems, some have argued in favor of a different approach. "In the presence of such deep uncertainty, the machinery of prediction and decision making seizes up. Traditional analytical approaches gravitate to the well-understood parts of the challenge and shy away from the rest. . . . Rather than seeking to eliminate uncertainty, we highlight it and then find ways to manage it" (Popper, Lempert, and Bankes 2005, 66).

When we cannot know how likely a future state will be, whether it be an increase in the surface temperature of the earth, or the rate of inflation, or the number of species going extinct, there are a number of different attitudes we might take. People vary in their approach to uncertain situations: from optimism to pessimism, from thinking that technological innovation will solve the energy crisis and avert the dire consequences of global warming to thinking it is already too late to slow the inevitable destruction of the planet, from best-case scenario to worst-case scenario. If we knew how likely the desired and undesired outcomes are with respect to various courses of action, we could use the calculus developed for decisions under risk to make a reasoned judgment. When these probabilities are themselves very uncertain, we fall back on our more basic attitudes toward uncertainty and intuitions. These are not good heuristics for rational action (see Morgan and Henrion 1990).

How should we think about an uncertain future? Considering only a single outcome projection—best case or worst case—as a foundation for decision making is not necessarily the best way to proceed. Pretending that all possible outcomes are equally likely flies in the face of what we do know about the future trajectories of complex systems. There are alternatives. "Scenario analyses" that consider multiple possible outcomes, even though the likelihood of one or the other is unknown, provide a way to hedge one's bets against an uncertain future. Various alternative strategies, which recognize the need to deal with the complexity and dynamic nature of the world, are emerging. One example is the methodology called "robust adaptive planning" (RAP), developed by Robert Lempert, Steven Popper, and Steven Banks (Bankes, Lempert, and Popper 2001; Lempert, Popper, and Bankes 2002; Popper, Lempert, and Bankes 2005). The four foundations of RAP are

- "considering large ensembles of scenarios,"
- "using robustness criteria to compare strategies,"
- "employing adaptivity to achieve robustness," and
- "using the computer as a tool to alternatively suggest breaking scenarios for candidate strategies and clever hedging actions to improve those strategies" (Lempert, Popper, and Bankes 2002, 438).

Rather than maximize expected utility, Popper, Lempert, and Bankes recommend identifying and adopting what they call the most robust strategies. These strategies might not have the best possible option available as any one outcome, but their satisfactory outcomes occur in the largest range of future contingencies. Robustness analysis requires one to consider models that take into account what we do know, without pretending that we have precise probability assignments for what we don't know. Rather it analyzes a range of diverse but possible scenarios and the ways in which a policy decision today would play out in each of them. As they put it: "A key insight from scenario-based planning is that multiple, highly differentiated views of the future can capture the information we have about the future better than any single best estimate" (Lempert, Popper, and Bankes 2002, 423). RAP is consistent with traditional mathematical methods for modeling and assisting decision making. However, it accommodates the realities of deep uncertainty and complexity that show that traditional techniques by themselves won't work. Deep uncertainty is simply the condition discussed above: a situation in which a probability (or associated measure) cannot be assigned to the influence of variables and, correspondingly, cannot be assigned to alternative outcomes of different actions. Conceptually, RAP depends on creating a large number of simulations in which many, many scenarios are created, each representing a different distribution of probability assignments to the relevant variables, their influence, and therefore the expected outcomes. This approach takes advantage of the fact that modern computing capacity provides the ability to analyze the nature of a large number of scenarios to provide us information about

- which scenarios are least sensitive to the uncertainties, so that we are more likely to get desired outcomes, that is, which are most robust; and
- in what ways the most robust scenarios might be "defeated" by unexpected changes or "surprises."

By assessing the performance of a number of different policy options in a range of plausible scenarios of the future, the analyst can see where policies would fail, and have the opportunity to craft new hybrid strategies that might be more robust than any of the ones initially considered. Having found a robust policy, one then proposes new scenarios that would make it fail and tries to build adjustments into the policy to overcome the undesirable outcomes in those situations. Of course, the human imagination can generate scenarios that, while plausible, may not be very probable. Lempert, Popper, and Bankes have proposed ways of accommodating the knowledge we do have of the likelihood of different scenarios to see whether or not one that "breaks" a robust policy should worry us or not.

An application of scenario analysis was developed for comparing policies for limiting greenhouse gas emissions. Scientific evidence supports the claim that burning fossil fuels increases the concentration of these gases in the atmosphere and that over the last hundred years this has altered the earth's climate system (Oreskes 2004). However, there is deep uncertainty about the future costs and benefits of alternative policies to address this problem. Even if we agree that we need to reduce emissions by some amount, what is the best way to do this? Technological innovation will be an important component in reducing emissions by as much as 80% from current levels, yet which policy taken today will better ensure that new technologies will be adopted and hence will be better at reducing emissions? Lempert (2002) compared two policies for their effect on technology diffusion: carbon tax (limits only) versus carbon tax plus technology subsidy (combined strategy).

> The agent model of the effects of the alternative policies on the climate and economic systems presents deep uncertainty in over 30 different input parameters that represent a wide range of factors, including the macroeconomic effects of alternative policies on economic growth, the microeconomic preferences economic actors use to weigh the cost and performance of alternative technologies, the characteristics of new technologies, and the dynamics of the climate system. (Lempert 2002, 7311)

The models that are projected into the future give forecasts of expected emissions that differ by more than an order of magnitude. The scenario-analysis simulation evaluated the policy alternatives against plausible futures characterized by a combination of features including the amount of greenhouse gas emissions after fifty years and the rate of cost reduction for nonemitting technologies, as well as the rate by which agents learn about performance, agents' risk aversion, and the heterogeneity of price-preference for new technologies. The tax-only option turns out to be preferable over the combined tax-plus-subsidy option in futures in which agent technology preference is homogeneous and the damage due to climate change emerges slowly. The combined tax-plus-subsidy is preferable when those conditions do not hold. Overall, the combined strategy dominates in a robust region of results even where there is only a modest expectation that climate change damage will be significant. This type of analysis permits the knowledge we do have of current factors and options to guide decisions when there is deep uncertainty about the probability of predicted outcomes.

In addition to new forms of representations and analyses where robustness with respect to future outcomes replaces maximization as the mark of a good policy, there are alternative policy heuristics that can help manage decision making in a complex world. Indeed RAP incorporates "adaptive management" as the appropriate vehicle for implementing the results of multiple scenario analysis. Adaptive management in particular recognizes not just the uncertainty of our knowledge of, say, the behavior of the ecosystem ten or twenty or thirty years from now, but is designed to take into account the increasing and changing knowledge of complex systems as the implementation of a policy is carried out (see Oglethorpe 2002; Buck et al. 2001).

Adaptive management is an action strategy most widely discussed in ecosystem management. Two scientists who have written seminal papers on adaptive management summarize the approach as follows:

> Adaptive management is a systematic approach for improving environmental management by learning from management outcomes. We believe that protected areas management can benefit greatly from this approach which allows management to proceed despite uncertainty, and reduces this uncertainty through a systematic process for learning. We describe this approach as a six-stage process: problem assessment, experimental design, implementation, monitoring, evaluation and management adjustment. . . . [A]daptive management is more than just a procedure; it also requires curiosity, innovation, courage to admit uncertainty, and a commitment to learning. (Murray and Marmorek 2003, 1)

There are many critical aspects to adaptive management. Perhaps the most difficult to achieve are the linked activities of experimental design, monitoring, and evaluation. The basic idea is common sense and one that has been practiced in organization management for years: implement a strategy but set short-term, measurable milestones so that you can evaluate whether the strategy is working or not. This approach recognizes that the world is not the one Laplace described: one in which the current state determines all future states. Nor can there be prediction of all future and past states from knowledge of all the forces and all the positions of matter at the present. A Laplacean world, governed by Newtonian-style laws, would certainly warrant a predict-and-act strategy. Our complex, feedback-rich, contextually contingent world requires that we intervene based on the best possible information, specify in some detail how to measure whether our expectations for the interventions are coming true in the short term, and modify our intervention strategy based on the feedback and corresponding adjustment of our hypotheses about the intervention's effect. It not as simple or easy as we would prefer; neither is the world we are trying to manage.

For the purposes of this discussion, the most important thing to notice about adaptive management is that it modifies the predict-and-act model to be an iterative process of predict, act, establish metrics of successful action, gather data about consequences, predict anew, establish metrics of successful action, act, gather data about consequences, predict anew. . . . Adaptive management is a dynamic, iterative, feedback-rich strategy for decision making that matches the dynamic, feedback-dependent reality of complex systems.

The uncertainty of knowledge about consequences of complex ecosystems is due to the presence of unknown effects of the multitude of causal components, the presence and effects of currently unknown components (see Myers 1979), and the uncertainty of chaotic dynamic processes. As was discussed above, this makes assigning a precise quantitative probability to a future state, like global warming, virtually impossible if we insist on a predict-and-act strategy, thus leaving our decisions insensitive to the types of knowledge we actually possess (see Clark et al. 2001). Just as both robust scenario analysis and adaptive management provide alternatives to an all-or-nothing strategy by requiring monitoring and updating as new knowledge is acquired, so too may a more nuanced representation of consequences of policy, one more sensitive to context, contribute to improved and more nuanced strategies for action in a complex world.

The Case of Genetically Modified Food

Let's consider the case of policy concerning genetically modified (GM) food in terms of the character of the uncertainty that arises from the complexity of the biological world. What are we trying to accomplish with a policy concerning genetic modifications of plants and animals?[2] There may be a number of different and sometimes competing values that we want to pursue, including economic value, the increased quality of health, the protection of the environment for future generations, the elimination of poverty and starvation in the world. Which effects of introducing GM foods or banning GM foods contribute to or diminish the values we are pursuing? To answer this we need to know the facts of the matter. What will happen if we increase the agricultural yield of rice on poor soils, or introduce biologically fortified "golden rice" that carries with it nutrients that are normally available only in a more diverse diet? What is the environmental effect of growing crops that have built-in resistance to pests that are otherwise managed by chemical pesticides? What are the consequences if we don't do these things?

Recall the traditional predict-and-act logic. If we know what we want (our values) and we know what we get if we adopt different actions (the facts and governing causal relations), then we are in a position to know what we should do, how we should proceed. Do a cost-benefit analysis and maximize expected value. It seems pretty simple. However, this simple account of a policy decision masks two sorts of complexities that muddy the waters. Not surprisingly, the two sources correspond to the two types of input. That is, sources of uncertainty for policy derive from, first, the complexity of biological systems and, second, the pluralism of values held by agents for whom and by whom the policy is made. The complexity of the causal structures of biological systems precludes simple sound-bite accounts of the consequences of genetic interventions. For example, the most widely used transgenic pest-protected plants, created by taking a gene from another organism, express insecticidal proteins derived from the bacterium *Bacillus thuringiensis* (Bt). Including types of corn, potatoes, and cotton, "Bt GM organisms" are often thought of as the result of a single kind of intervention. But Bt toxins are not all the same; they vary genetically and biochemically (National Academy Press 2000, 109; Marra, Pardey, and Alston 2003). Furthermore, the consequences of the use of Bt plants vary in their environmental impact on other organisms and wild types and with respect to their pesticide-reducing benefits. Replacing regular corn with Bt corn provides insect resistance, but does not extensively modify the use of externally applied chemicals because these are generally not used for the job Bt does for corn (Wu 2006). The use of Bt potatoes, however, does reduce the amount of pesticide required for healthy crops (National Academy Press 2000, 111). So what sort of policy should be adopted for Bt plants in general?

Bt potatoes are successful at resisting a variety of beetles. Their use can reduce the need for all the externally applied chemical insecticide as currently used for that purpose. Yet the genetic modification may also have consequences for nontarget insects, or induce an escalation in adaptations to evolve nonresistant parasites, or have pleiotropic effects on the modified plant itself or the landscape of genetic diversity in the wild, and so on and so forth. The waves of consequent effects are not only broad, but at least some of the interactions with intricately connected parts of the ecosystem are still unknown. Since the consequences of the same genomic strategy for different agricultural plants are different, say, for Bt corn and Bt potatoes, even if we understood all the implications in the case of Bt potatoes, these would not necessarily apply directly to cases of Bt corn. The context sensitivity of the same kind of intervention in biological systems makes the challenge of rational decision making for GM foods, discussed earlier in this section, all the more challenging.

Since science is an ongoing process of discovery, our ongoing assumption should be that we will need routinely to update our policy relative to what we discover. This is in stark contrast to other environmental policy commitments, like the "no surprises" clause in the Multiple Species Conservation Program that was implemented in San Diego County and heralded as a model for the entire country (Jasny, Reynolds, and Notthoff 1997). There the rule is that, once an agreement is made between local government and developers on what lands are to be set aside, nothing that may be discovered in the future can overturn that decision. This type of policy, though perhaps optimal from the developers' point of view, does not reflect the type of scientific process involved in understanding endangered and threatened species and their habitats. Rather the current uncertainty combined with the incremental and fallible knowledge gained with continued scientific investigation requires us to apply adaptive management (Mangel et al. 1996). We need to update, overturn, and amend particular decisions as our knowledge of the consequences changes.

For the "unknown unknowns" (Myers 1995), if there is reason to suspect that the environmental harm will be great and irreversible, then we are facing a situation of deep uncertainty, and scenario analysis, as suggested above, may take better account of what we know and what we do not know in providing a tool for us to act reasonably and responsibly. More typically, deep uncertainty has invited precaution. A precautionary stance in article 15 of the Biodiversity Convention states: "In order to protect the environment, the precautionary approach shall be widely applied by states according to their capabilities. Where there are threats of serious or irreversible damage, lack of full scientific certainty shall not be used as a reason for postponing cost-effective measures to prevent environmental degradation" (http://habitat.igc.org/agenda21/rio-dec.html). Scenario-analysis methodologies, like RAP, provide one interpretation of how to take a precautionary approach.

The precautionary principle is lauded by environmentalists who fear the possible, though unproven, damaging effects on the environment of new technologies. It is simultaneously dismissed by those who take it as an unjustified prohibition to innovation (see Cranor 2001; Foster, Vecchia, and Repachoili 2000; Bodansky 1991; O'Riordan and Cameron 1994; Kaiser 1997; Lacey 1999, 2002; Smith 2000).

The Precautionary Principle

The original intent of the precautionary principle was to shift the legal burden of proof from regulators having to prove with "full scientific certainty" that harm would be done in order to ban a procedure, to those who want to introduce some new technology. Under the principle the innovators might have their products prohibited unless there is "full scientific certainty" that no harm could be done. Interpreting the principle with an emphasis on the proof of harm, or proof of no harm, invokes a type of epistemological positivism that is entirely unwarranted. Given the nature of the complexity that I have described, there can be no scientific certainty either way. Rather than pass the burden of unattainable "proof" from one party to the other, we must acknowledge and manage the inescapable uncertainty. By not adopting some transgenic practices that may not induce irreversible harm, we are at risk of not reaping the benefits of those technologies out of unjustifiable fear—just as by adopting a new technology that does issue in irreversible harm, we are risking the harmful consequence. A precautionary approach is indicated by the nature of the possible harm. If the possible harmful consequence is, as stated in the principle, both serious and irreversible, then clearly more caution is required than if it is minor and reversible. But caution here has to reflect the nature of the uncertainty. It does not entail inaction. Indeed, inaction takes science out of the decision process. Rather, complexity suggests a more flexible policy response than total ban or complete hands-off nonregulation.

How then can policy procedures take science seriously, when the science invokes perhaps deep uncertainty with respect to future consequences of new technologies? A fixed policy—ban GM foods for all time—is an inappropriate response to a dynamically growing scientific base of understanding. If our policies are to reflect what we know about the world, as they should, and our knowledge is changing relatively rapidly, then our means for policy amendment should be sensitive to those dynamics. If our policies are not so constructed, we will be stuck with the problem of having to decide once and for all on the basis of information that is ephemeral. The alternative to a flexible management response, given the impermanence of our understanding, is to have policy ignore science altogether. Since there is no univocal accounting of the precise consequences of genetic interventions on health, productivity, biodiversity, and so on, a silencing of scientific input into policy is all too easy. This would be a very bad thing. Uncertain knowledge is still better than ignorance.

Uncertainties about the consequences of the known variables in a complex ecosystem suggest a policy procedure that endorses continued investigation and is flexible enough to take into account our changing knowledge and adjust our actions appropriately. This will require close monitoring of new interventions with ongoing feedback from scientific analyses. Monitoring itself is not easy, especially given the variety of genomic interventions and variability of local environmental impact (Diamond 1999). What is true for Bt potatoes is not true for Bt corn.

Other approaches to managing uncertainty in environmental policy have been proposed. However, I believe they are inadequate to address the special circumstances of genetic modifications with potentially irreversible effects. For example, Robert Costanza and Laura Cornwell (1992) suggested implementing what they call the 4P approach to scientific uncertainty. The four Ps stand for "precautionary polluter pays principle." The suggestion is to introduce an assurance bonding system that requires those who want to introduce a potentially harmful new product to make a commitment of resources up front to offset the potentially catastrophic future effects. It is a flexible system insofar as portions of the bond would be returned if and when it can be demonstrated that the suspected worst-case damage has not occurred.

This economic approach shifts both the burden of proof and the cost of uncertainty from the general public to the agent who wants to introduce the new technology, bypassing a need for specific safety standards to be established. However, it seems to me that while this might work for harmful human health consequences of introducing a new agent into the environment, it is not clearly appropriate for the type of environmental consequences for biological diversity and ecosystem health at stake in the case of GM foods. That is, for harmful human health effects of a new product, perhaps the innovator can compensate the humans whose health is damaged. But how much money would be required to offset the irreversible, cascading effects on the ecosystem? Rather than go full steam ahead and just be prepared to pay up if there is harm, I think a slower, monitored pace with the option to stop if the harm is looking more probable is better fit to this situation. Thus where the harm is serious and irreversible, and where there is no current procedure for reducing the uncertainty, an iterated, adaptive approach that analyzes plausible scenarios, attaching probabilities when and where we have such knowledge, is better able to promote our values in a dynamic, changing, and complex world.

In this chapter I outlined challenges that derive from the many forms of natural complexity for our making rational policy decisions to promote our interests. The many forms of complexity can produce ineliminable uncertainty and subvert a predict-and-act approach based on assumptions that the underlying science can always be described by facts and by governing causal relations. Although for some cases, reducing the uncertainty to a manageable quantitative risk may be feasible, for other cases, notably global climate change and technologies that impact biodiversity, holding out for such a resolution could be disastrous (see Sarewitz and Pielke 2000). Alternative perspectives on how to represent uncertainty and how to manage policies adaptively require a shift in expectations regarding the character of how scientific knowledge of complex systems can be implemented in policy analysis and decision making.

How, then, do we deal with these types of factual uncertainty in making sensible policy decisions? We have explored some specific techniques such as robust scenario analysis and adaptive management. What can we say more generally? Policymakers should

- acknowledge and manage the known risks;
- continue to investigate the unknown consequences on known factors; and
- adopt robust adaptive planning, adaptive management, and other such scenario-based, feedback-based, metric-driven approaches to deciding in the face of deep uncertainty and complex systems.

Policymakers would like neat, certain answers to questions of risk, so that an easily enforceable policy can be made. For the reasons given above, however, fixed probability assignments and law-governed predict-and-act models cannot reflect our scientific knowledge in many of these situations. We cannot pretend that there is certainty when there is not—and we cannot hold out for certainty when it is not going to be found. In an interview in March 2000, Edwin Rhodes, aquaculture coordinator for the National Marine Fisheries Service, said he was surprised to hear that the Food and Drug Administration was overseeing the environmental review regarding the new GM salmon into which a foreign gene was introduced that keeps its growth hormone continually rather than cyclically operant. Rhodes said the National Marine Fisheries Service, not the Food and Drug Administration, had the expertise to make decisions on such things as whether GM fish should be grown in net pens. "We have to have absolute certainty that transgenic fish do not interact with wild stocks," Rhodes said (Yoon 2000). We will never have certainty—to make our policy depend on it is a mistake.

How might this set of concerns address the issues raised in our initial case of complexity, namely, the role of genetics in major depressive disorder? Funding policy is an area where both uncertainty and adaptive strategies might be considered. The first concerns research decisions by granting agencies. Given that there are many different causal contributions to generating major depressive disorder, and their specific contributions and interactions are not well known, a bet-hedging strategy might be appropriate for distributing funds to different areas of research. Given the state of uncertainty regarding genes, environmental influences, and other components of the complex etiology of adult depression, and given the aim of developing effective treatments, it is at least unwise to put all one's eggs in a single basket. Funding only one of the different types of investigation—gene research, neurochemical research, family psychological research, and so forth—would be unwise. Decision makers might want to take an adaptive stance toward this problem, engaging multiple "learn by doing" options with monitoring reviews to redirect funds toward projects with more promising methods or more successful results.

6

INTEGRATIVE PLURALISM

This book set out to display some of the challenges that the scientific study of complex natural systems raises for traditional approaches to epistemology and to suggest a way forward. To begin to understand our complex world, I have argued that we need to expand our conceptual frameworks to accommodate emergence, contingency, dynamic robustness, and deep uncertainty. Our conception of the nature of nature must shift away from the expectation of always finding regularities and causal powers that are universal, deterministic, and predictable. The truths that attach to our world are rarely simple, global, and necessary. Rather, nature organizes itself in a plurality of ways, and what we validate as "knowledge" should reflect that diversity. Adequate epistemologies will capture the context richness and degrees of contingency of causal processes in general truths that do not apply to all time and space, but only to more limited periods and regions in a dynamic and evolving universe.

Complexity reveals the historical character of events as they unfold, the number of components contributing to the structure and behavior of natural systems, and the robust, flexible responsiveness of many natural systems to internal and external changes. Complex systems defy simple methods of investigation and simple logics of inference. The expanded epistemology I advocate does not jettison traditional epistemological approaches for those contexts in which they work; it promotes instead a more balanced, pragmatic approach necessitated by the realities of the world we are trying to understand. We do not live in Heraclitus's ever-changing and unknowable world of eternal flux. But neither do we live in Parmenides' world of the unchanging, where only the eternal is knowable (see Kirk, Raven, and Schofield 1983). Instead we live in a world that comes in many shapes and sizes with structures differing in degrees of stability, affording more or less contingent truths that we can know and use to pursue our goals and aspirations.

Integrative pluralism, the view of science that I have presented, has implications for the ways in which we craft epistemology and policy. It takes the lessons from studying the sciences of the complex to conclude that, in many domains, only nonreductive, *integrated* multilevel, pragmatically targeted explanations will succeed. Explanations that take into account the complexity of the world will be, almost by definition, more local and less global to reflect the context richness of the behavior of multilevel, multicomponent systems. In each area of science and life, how what we have designated as context shares in the causal responsibility for generating effects will vary. How much or how little specific components, which occur at many different levels, affect system behavior will depend on the stability of the contingent causal structures. For example, the phenotypic upshot of a single genetic mutation will, in some cases, vary with differing internal and external backgrounds. Recall the case that opened the discussion. The genetic variant for a short promoter region of 5-*HTT* contributes to the susceptibility of adults to major depressive disorder, but only when it is expressed in the context of traumatic experiences, which thus also contribute to the incidence of depression. These experiences act as triggers in a feedback system that is more sensitive for the short-allele individual, and thus she feels stress at lower thresholds than those with the long allele. This feeling of stress in turn creates even higher sensitivity to the potentially traumatic features of the environment, creating more stress, and so on. In such situations the individual often retreats from or suffers from amplified traumatic experiences. Studies on nonhumans have supported this hypothesis, showing how maternal behavior modifies the experience of infants by moderating the number of stressful events they are likely to experience. Multiple factors contribute to the stress response cycle in the studies of nonhuman animals that can amplify into what is comparable to human adult depression (Pryce et al. 2002).

Reductive and exceptionless generalizing approaches will be successful in some situations, for example, the relationship between mass and gravitational force and between electric charge and electromagnetic forces. Though not logically necessary, many regularities are nevertheless extremely stable, and thus for the most part impervious to the changing seas of other conditions in which they sail. In other situations, as with the etiology of psychiatric conditions like major depressive disorder, a reductive genetic explanatory method will fail. As I have reported, for that case the basic genetic components of the complex structure of cells, brain, and organism do have a predictive efficacy for adult depression but only when coupled with experiences of childhood trauma. And correlatively, experiences of childhood trauma predict adult depression only in the presence of one or two representatives of the short-allele genetic factor that helps build the serotonin receptor structure in the brain. Genes play a role, but not the dominant or unique explanatory role a simple reductive strategy would expect.

Is reduction always wrong? No. Is reduction always right? No. Reductionism becomes the enemy only when it is promoted as the only game in town. Pluralism in explanatory strategies recognizes the diversity of causal structures that populate our world. But what kind of pluralism? There has been much discussion of disunity or pluralism in philosophy of science (Galison and Stump 1996; Kellert, Longino, and Waters 2006); however, not all versions of pluralism are alike. There is the "anything goes" pluralism attributed to Paul Feyerabend (1975, 1981), which follows from nihilism about effective investigatory methodologies. What was the source of that nihilism? Kuhn's enormously influential book *The Structure of Scientific Revolutions* (1962) exposed the fact that in the history of science the very same methods delivered both progressive results and utter failures. Kuhn's book both summarized and expanded on the observations made by a number of mid-twentieth-century philosophers who pointed out the epistemic complexities of *the* scientific method (Hanson 1958; Lakatos and Musgrave 1970; Gower 1997), including the theory-ladenness of observation claims, the inability to measure simplicity or fruitfulness, and the like. Since then, trust in articulating *the* scientific method leading to *the* truth about nature has waned. Indeed, Kuhn even argued that there was no necessary connection between the truth of a theory and fulfilling any or all of the values that science constitutively endorses for preferring one hypothesis to another (i.e., simplicity, fruitfulness, accuracy, consistency, and scope). For some sociologists of science (Collins 1982, 1985; Latour 1988; Woolgar 1988), these results led to the replacement of truth as the aim of science and the explanation for why one theory is and should be preferred over another, with authority and power in the social institutions of science. But while a survey of contemporary scientific practices will indeed yield a vision of a plurality of theories and models, increasing rather than decreasing in number as science grows, I do not believe any actual scientific practice supports "anything goes" pluralism or provides reason to abandon the belief in well-confirmed scientific theories as representative of the world. Empirical test remains the arbiter of scientific worth, and though experiments invoke many assumptions and values beyond the ones at test, acceptance of what "works" is not determined exclusively by who is editor of the premier journals or who is getting the biggest grants.

There are more moderate pluralist options available. Some (Friedman 1974; Kitcher 1981; Beatty 1987) recognize and promote a temporary plurality of competing theories as a means toward achieving unity of science in the long run. Competition is endorsed as a way to hasten the journey to acceptance of the best unified schema that can explain the greatest amount of diverse data. Given our uncertainty regarding which of two or more exclusive options may be correct, say whether humans are more directly related to the chimpanzee or to the orangutan (see Schwartz et al. 1984; Schwartz 2004), this view of pluralism argues that it is rational for the community of scientists to entertain them both, until empirical evidence provides overwhelming support of one over the others. As Kitcher (1991, 19) says, "The community goal is to arrive at universal acceptance of the true theory." Some of the diversity of theories and explanations that are found in science will yield to this strategy, but not all. In the end, this is a pluralist methodology for achieving a monist goal: achieving a single, true unified theory via competition of exclusive alternatives. In complex natural phenomena, it is often the case that only parts of the array of cotemporaneous scientific claims are in competition; other parts are compatible and, indeed, must be integrated into a multilevel explanation for explanation to succeed.

Integrated Multilevel Explanation

The ontology of complex systems suggests a different form of pluralism. The multilevel structure of complex systems encourages focused analysis of causal structures at each level. Yet these explanations—say of human behavior in terms of genes, or hormones or memories from childhood or socioeconomic milieu—do not clearly vie for the spot of "one true explanation." Competition among mutually exclusive theories describes one aspect of the scientific landscape, but it does not tell the whole story. Individual explanations making up the variety of explanations even for a single phenomenon, like major depressive disorder, are not always competing; they are sometimes compatible and complementary. In addition to the explanation of the genetic component of depression, there is also analysis at higher levels of organization. The monoamine hypothesis characterizes the causal role of neurotransmitters serotonin, norepinephrine, and dopamine (see Delgado 2000). There are studies of the relationship between depression and enlarged amygdala and reduced hippocampus regions of the brain (Weniger, Lange, and Irle 2006). And there are hypotheses identifying social and experiential causes (Lorant et al. 2007). Depression, typical of many psychiatric disorders, is an exemplar of a multicomponent, multilevel complex behavior.

As the causal structure underlying the generation of such behaviors is better understood, a methodology tuned to detecting the types of contingencies and interacting factors is more likely to yield usable knowledge. Although it would be much easier if complex behaviors were the result of the presence of a single structural allele that is expressed in all contexts, that is not what is found. Emergent systemic properties at higher levels result from interactions of lower-level components, and those components are responsive to internal and external context. The lower-level components, in many cases, also interact with the emergent properties. Scientific explanation that attempts to render only the most basic properties of the most basic entities as causally and explanatorily sufficient will miss the causes at higher levels that are an essential part of the story of complex systems. The relationships between factors at the various levels are not independent of each other: the analyses of each must be integrated with results from study of the others to determine the roles they play in generating the behavior of interest of the complex system. Indeed, there may be more than one set of interacting features at different levels. As Kendler, Gardner, and Prescott (2006) argued, there are multiple distinct pathways to a depressive episode tracking through internal, external, and social factors over the lifetime of an individual. Not only is there no single, reductive factor that will explain why a person suffers from depression; there is no unique composition of causes or explanation involving different levels that will do the job for all cases. Integrating information about the genes, cells, brain, chemistry, affect, and environment will not be algorithmic. Pluralism in causes, pluralism in ontological levels, and pluralism in integrations will characterize the scientific explanation of complex system behavior.

The historical character of evolved complex systems is another source of multiple explanations and a de facto pluralism. Why do some insects cohabit and have division of labor among individuals and others do not? One answer comes from an evolutionary perspective on the adaptive benefits social life gave to originally solitary insects and thus contributes to an account of how such a system has evolved independently multiple times in the history of life (Wilson 1981). Another answer points to the proximate mechanisms in which interactions among individual ants, bees, or termites cause division of labor in colonies today, and the particular set of hormonal, developmental, genetic, and self-organizing processes that produce social life (Topoff 1972; Camazine et al. 2001; Robinson 1992; Robinson, Grozinger, and Whitfield 2005).[1]

The biological questions one can ask of an organism's trait can be further partitioned to delineate more levels of analysis, including evolutionary origins, current reproductive function, ontogeny, and proximate mechanism (Tinbergen 1963; Sherman 1988; Beatty 1994; Mayr 1993; Sterelny 1996). Clearly asking different questions of some part of nature will yield answers that are not directly competing with each other. It is like asking what the tallest building in the world is and asking how the building was constructed to be that tall. There may be disputes about which building is the tallest (should you count towers and spires or not?), but whatever the answer is to that question, it cannot be contradicted by knowing the construction techniques that were used to build it.

Granted that there can be different answers to different questions, it still remains to be determined what relationship holds among the answers. Are they consistent, independent, connected in any way? There will be causal connections between a building's being one of the world's tallest in 2007 and the construction techniques used. None of those buildings will be made of wood; all will have mechanisms to deal with sway effects of high winds. Sherman (1988) posited that competition between questions could only occur *within* a level of analysis and not between them. Consider an evolutionary question, namely, the origin of sterility in social insects, an important aspect of division of labor where only the queen produces offspring, while the workers suppress their own reproduction to take care of the hive and their sisters. Natural selection operating on individual genes by means of kin selection was offered by Hamilton (1964) as an explanation to answer the evolutionary-origins question. His kin-selection explanation was based on the unusual haploid-diploid genetic structure found in social insects. Because females develop from fertilized eggs and males from unfertilized eggs, worker bee females are more closely related to their sisters (the offspring of the queen) than they would be to their own daughters and sons, if they had them, and hence a worker gets more of her genes into the next generation of bees by raising sisters. This explanation competes with an evolutionary explanation of worker sterility by means of selection on individual queens for their ability to control their daughters. No special kin relationship is required for queens to be selected over other queens if they are successful at becoming parasitic on the workers they control (Michener and Brothers 1974; Andersson 1984; for a comparison of the competing theories, see Alonso and Schuck-Paim 2002). Get the workers to, well, do the work and take the risks of going out in the world to forage, and as a queen you will be more reproductively successful than a solitary female having to do all the tasks from building a nest to provisioning it with food to laying the eggs and fighting off predators. At the level of proximate mechanism, one can ask how it is that a specific female bee becomes a queen as opposed to a worker, and how, once assigned a role, their reproductive abilities are enabled or suppressed. According to the levels-of-analysis model of the plurality of scientific hypotheses and explanations, answers to the mechanism questions should not compete with answers to the evolutionary-origins question. Yet it would be wrong to think the answers are completely unrelated to each other.

An extreme interpretation of the levels-of-analysis method of partitioning of questions and their corresponding answers could lead to a form of isolationism where scientists working on the same phenomena, but focusing on different levels of analysis, would have no need to consider the explanations developed at levels other than their own. While the levels-of-analysis approach correctly recognizes the diversity of questions that can be raised, it fails to acknowledge that the answers at one level may well influence what can be a plausible or probable answer at another.

Returning to the division of labor in social insects, recent studies of how workers' reproduction is proximately controlled, and how workers who have developed ovaries are affected by being introduced into a colony with an active queen, have shown that the queen's pheromonal control over ovarian development is strong and direct. Indeed, workers who in isolation develop ovaries will reverse that development when a queen is introduced, even though the worker is ready to lay an egg and would increase her individual fitness by so doing (Malka et al. 2007). This lends support to the queen-control "ultimate" or "evolutionary" explanation of sterility in insects and raises questions for the completeness of the kin-selection evolutionary explanation. Answers to the proximate questions can describe and explain the phenotypic traits that occur by opening the black box to investigate the mechanisms that generate them. How it in fact is done can constrain the range of possible variations that are likely to have occurred in the past, and thus speaks directly to the assumptions underlying evolutionary ultimate explanations of the origin of the trait.

Evolutionary-developmental biology, or "evo-devo," recently emerged as a new field in biology that explicitly addresses the relationship between developmental processes responsible for generating the traits of an individual organism, evolutionary processes responsible for the selection and maintenance of the traits, and the mechanisms for generating them in populations over history (Oyama, Griffiths, and Gray 2001).

> Development may make a crucial contribution to evolutionary theory. Modern evolutionary biology has focused on the role of natural selection, which operates external to the organism, and views organisms as unconstrained in variation. Micro-evolutionary processes are considered sufficient to explain macro-evolutionary history. However, developmental processes are emergent, and not predictable from the properties of genes or cells; therefore, starting with a particular ontogeny, some phenotypes might be readily achieved and others impossible. Developmental mechanisms are crucial, both to large-scale evolutionary changes, and also to small-scale evolutionary processes. (Raff 2000, 78)

Yet another source supporting the relevance of answers to questions raised in variant levels of analysis is studies of different self-organization models that show that some emergent features of a social insect colony arise "for free" just by the interactions of the individuals (see Page and Mitchell 1998; Camazine et al. 2001; Bonabeau et al. 1997).[2] By appealing to natural selection, evolutionary explanations require ancestral variation in a trait. The possibility that natural selection is the major source of the prevalence of a trait in a population can be challenged by proximate explanations that give evidence for the implausibility of variation having occurred in the past. Results at one level influence the answers to questions at another. Isolationism with respect to levels of analysis cannot be endorsed.

Even within a level, where Sherman (1988) suggests competition among explanations should properly occur, the framework breaks down. The levels-of-analysis détente strategy operates by partitioning the questions into four different levels: evolutionary origin, current reproductive function, ontogeny, and proximate mechanism. If questions address different levels, there is no opportunity for competition between answers to them, and if they are in the same level, then competition can occur. But this does not strictly work as a description of the relationships between different parts of biological research. The multicomponent causal structure of complex systems challenges this popular framework.[3] When there are multiple causal factors at work—like gravity and friction, or genes, environment, and learning—it is common practice to model each factor individually, to understand for what part of the effect a contributing cause might be responsible. In such cases the same question (e.g., what are the self-organizing factors generating division of labor) might generate different answers that are focused on partial contributing causes of the phenomenon (Mitchell 2000). But although these separate causes might be fully controlling the effects of a system, generally they contribute only partial answers. Thus the models of individual causes will often need to be integrated to explain particular occurrences of the phenomenon. This is clearly the case for the multiple causes at different levels of organization for major depressive disorder.

The view of pluralism that I endorse is not "anything goes" or "winner takes all" or "levels of analysis" but rather *integrative pluralism*, which attempts to do justice to the multilevel, multicomponent, evolved character of complex systems. But, one may reasonably ask, what kind of integration? From the examples above, both of major depressive disorder and of division of labor in insects, it should be clear that there is no algorithmic method for combining different contributing causes and different levels of causation for the types of complex systems I have been discussing. For some component causes, simple aggregative rules will work. For example, vector addition in physics is a general method for combining the effects of independent forces of, say, magnetic and electric forces on the motion of a body. Since these are additive components, a combinatorial rule will give the resultant total force and direction of the body. It has been argued (Sober 1987) that this type of integration may be adequate to predicting the joint effects of mutation and selection on gene frequencies in a population. But given the context sensitivity and variety of not just the components structuring complex natural systems but the *ways* in which they interact in generating a system property, such simple rules are unlikely to be widely applicable. Even in the case of self-organization of division of labor, where genetics, learning, and the environment of the individual insects may all contribute to how the colony sorts its individuals into various tasks and specializations, the ways in which they contribute will vary from one species to the next.

I have provided some examples of how multiple explanatory factors operating at different ontological levels enter into explanation in the biological and psychological sciences. I cannot generalize the details of these cases. Indeed, one of the major claims I am making is that how a diversity of factors and the models of their behavior are combined into an integrated explanation is itself context sensitive. Unfortunately, this leaves a picture of science as having substantially greater challenges than the one painted by a thoroughgoing reductionist approach. On my view, there is no privileged level to which all explanations must be directed; there are multiple levels of causes. The reality of the successes and failures in contemporary sciences of explanations of complex natural systems allows us no other conclusion. The challenge for both scientists and philosophers of science will be to specify and identify the scope of principles and methods of integration for the particular complex behaviors that are the target of explanation. Integrative pluralism defines what Lakatos (1970) called a "research program" for both science and philosophy of science. It signals the end of a once noble reductionist quest but also outlines the beginning of efforts that will lead to a better understanding and more effective interventions in our complex world.

Pragmatic Considerations

While the nature of nature demands pluralism in the theories, models, and explanations we fashion, there is another impetus for pluralism arising from the way we humans come to know and represent nature. Scientific knowledge consists of *claims* about the causal structure of the world. Those claims are, inevitably, representations or models, whether they are linguistic, mathematical, visual, or computational. We choose at what level of abstraction we represent the features of the world we are attempting to explain. What determines if claims about the causes of major depressive disorder are accurate is ultimately the structure of the phenomenon, but what determines the level of abstraction of a representation, how fine-grain or coarse-grain are the descriptions onto which we map those structures, is a combination of our cognitive abilities and the purposes for which we intend to use the knowledge. There is the ontological reality, but there are also pragmatic choices about the representations we fashion to deal with that reality. Different representations and different levels of specificity work for different purposes.

Take an example from chemistry. There are a variety of molecular configurations that we commonly lump together as "water." Representing the variety at a finer grain is sometimes informative, sometimes not. We can, say for the purposes of understanding the effects of ozone depletion on the temperature of seawater, identify water coarsely, not discriminating the different configurations that the basic molecular form may take. Water is H_2O in the main form, but the hydrogen can be replaced with deuterium ("heavy water") or with tritium ("superheavy water"). While each of the three molecules has one proton and one electron, hydrogen has no neutrons, deuterium has one, and tritium has two. If we include the variations of isotopes of oxygen (O-17 and O-18), then there are, at a fine-grain level of representation, twelve different molecules that we can treat at a coarse-grain level as "water." Semiheavy water, HDO, occurs naturally in regular water at a proportion of about 1 molecule in 3,200, while heavy water, D_2O, is 1 part in 6,000. So for the most part, the ocean is H_2O, and there is no need to make the finer distinctions.

However, there are differences in the reaction rate and spectroscopic properties of H_2O and D_2O and HDO that are crucial to distinguish in other contexts. For example, proton nuclear magnetic resonance spectroscopy, used for ascertaining molecular structures, is often performed in deuterated solvents so that the only hydrogen in the sample will be from the molecule being analyzed (Patel et al. 1970). To accomplish the aims of this technology, not discriminating between H_2O, HDO, and D_2O would ensure failure. But for studying tidal properties of the ocean, or global-warming effects, representing the differences between H_2O, HDO, and D_2O, which are all present naturally, would just be confusing.

A literary detour makes the case vivid. Jorge Luis Borges's short story "On Exactitude in Science" (1998, 325) shows that the most precise representation may not be the most useful.

> In that Empire, the Art of Cartography attained such Perfection that the map of a single Province occupied the entirety of a City, and the map of the Empire, the entirety of a Province. In time, those Unconscionable Maps no longer satisfied, and the Cartographers Guilds struck a Map of the Empire whose size was that of the Empire, and which coincided point for point with it. The following Generations, who were not so fond of the Study of Cartography as their Forebears had been, saw that that

> vast Map was Useless, and not without some Pitilessness was it, that they delivered it up to the Inclemencies of Sun and Winters. In the Deserts of the West, still today, there are Tattered Ruins of that Map, inhabited by Animals and Beggars; in all the Land there is no other Relic of the Disciplines of Geography. J. A. Suárez Miranda, Viajes de varones prudentes, Libro IV, Cap. XLV, Lérida, 1658

When should we use fine-grain representations and when could we use coarser-grain representations? They are both "true" and "accurate" ways to capture features of the world we want to represent. This is the crucial point: what determines the level of abstraction or granularity is pragmatic and *can be answered only with reference to the particular context and scientific objective*. Different representations will be better for solving different problems. What we should expect in science is a diversity of representations, in addition to the plurality of hypotheses attaching to different causes and different levels of structure in complex systems.

Dynamic Perspectives

A significant lesson to be learned from the current sciences of complex systems is that dynamic processes responsible for the generation and maintenance of system-level properties challenge simple notions of cause and effect. With feedback interactions and chaotic dynamics, even a complete understanding of a deterministic structure in nature will not enable us to make secure predictions of the future. Add in quantum indeterminacy and computational intractability and it seems amazing that we can know anything at all about the complex world in which we live. But that amazement is founded on assumptions that knowledge requires that we can specify all the details of a Laplacean universe in which laws are universal and exceptionless, and causal structures are well behaved. If the assumptions entrenched in a Laplacean worldview define knowledge, then we cannot have knowledge of evolved, dynamic, robust, multicomponent, multilevel complex systems.

There is an alternative to despairing of the failure of the traditional epistemic strategies when applied to much of the biological world and probably most social phenomena to conform to the tidy picture of a simple world. However, it requires us to embrace more than the simple, and expect less than the fully predictable. I have offered integrative pluralism as the beginning of an expanded approach to epistemology, one that embraces diversity in nature, both dynamic stability and instability in causal processes, and ineliminable and deep uncertainty. To pretend that these do not exist in order to satisfy a preconceived notion of order in the world is to miss out entirely on what is in front of our eyes: a dynamically changing, complicated, complex, and chaotic but understandable universe.

Especially with respect to the impact of complexity on our plans and policies, holding out for something more tractable and more certain may well lead us into disaster. As I have argued, trying to squeeze various degrees of contingency into a rigid framework that has room only for universal truth or accidents, trying to force all robust causal feedback systems into well-behaved modular processes, and trying to pretend the deep uncertainty of some natural complexes will be replaced by consensual agreement on quantitative risk are to not face the facts of the world in which we live. The old Newtonian aspirations of reduction to the simple, most basic properties and most basic motions has been replaced by a world of multilevel causal interactions and emergence. A world of only necessary truths cannot engage productively with the degrees of contingent causation that characterize historically evolved natural systems. The universal has given way to the contextual and local, and a search for the one, singular, absolute truth must be replaced by humble respect for the plurality of truths that partially and pragmatically represent the world. Not only has our scientific understanding of nature changed, but science's methods of investigating nature have followed suit. Taking advantage of new computational techniques extends and transforms our ability to represent complex systems and study their behavior, so that we can now better engage with the details that matter (see Fox Keller 2002).

If we project the overly simplistic old views of science as *the* epistemology of science, then when simple explanations and methods fail in complex situations, it appears to policymakers that science fails. The danger is that holding science up to the wrong standard will diminish the value of what science discovers about nature, and could create an environment in which science is no longer consulted to inform policy. "The scientists can't give us a definitive answer, so why should we listen to them?" It is incumbent on us to develop and refine an understanding of the kinds of explanations that science can provide of multicomponent, multilevel, evolved, and robust complex systems, and correspondingly tune our decision making and policy strategies to that knowledge. Integrative pluralism is a step on the road to an expanded understanding of our complex world.

NOTES

CHAPTER ONE

1. Some have taken irreducibility to indicate phenomena inaccessible to naturalistic scientific explanation. However, that inference relies on an oversimplified account of explanation. Integrative pluralism, the position defended in this book, argues instead that simple reduction is not the only kind of scientific explanation. Analyzing complexity invites an expansion of our understanding of science, not a rejection of science.

2. Jane Kenyon's poem "Having It Out with Melancholy" (1996) expresses well the subjective feeling of depression.

3. The debate about whether it is nature or nurture that explains human behavior has continued to rage. For an account of the history of this debate, see Tabery 2008.

4. The Caspi et al. 2003 study has had a big impact on research on gene/environment interaction in behavioral genetics generally. However, a recent meta-analysis concludes that there is failure to replicate the original results for serotonin transporter gene interaction with environmental stress as predictive of depression. Risch et al. 2009 analyzed fourteen studies published from 2004 to 2009, six of which had confirmed the original study, at least partially. Yet their meta-analysis indicates no interaction effects and no 5-*HTT* gene effects on depression. They caution trusting nonreplicated studies. At the same time, the paper recognizes the difficulty in detecting weak causal effects of genes, environment, or their interaction, claiming that "characterization of gene-environment interaction has been most successful for diseases or traits that allow the study of a single gene with a major effect in the context of a relevant environmental exposure of varying magnitude, and also when the environmental exposure has a strong effect" (2469). Other types of studies—that is, on primates—have also suggested gene/environment interaction for depression. While this particular hypothesis might not be borne out in the long run, it is clear that the scientific study of complex behavior not only requires multiple forms of

investigation, but also is complicated when the causal contributions, though genuine, are weak either alone or in interaction. The research program I outline in this book does not depend on the success of any particular hypothesis, but rather reflects on a wide range of scientific practices.

5. See Murphy 2008 and Mitchell 2008b for alternative views of integrative strategies for psychiatric disorders.

6. See Schouten and de Jong 2001 for a thorough discussion of genetic reductionism in psychology and its alternatives.

7. This is not to say the multiplicity of causal components was ignored; indeed Sewall Wright (1931) invented path analysis in order to represent multiple latent variables. Rather, features that are typically not identified with a variable are often overlooked.

8. For further discussion of philosophy of science and psychiatry, see Kendler and Parnas 2008.

9. Since the 1970s, many scientific investigations have dealt with "complex systems" in the technical sense of systems that depend heavily on local feedback loops and are, as a result, extremely sensitive to initial conditions. I use the phrase "scientific complexity" to include this notion as well as other varieties of complexity that are also important in contemporary science.

10. See Lewontin 2003 for a discussion of simplification.

11. See Giere 1999 for a defense of a post-Enlightenment philosophy of science. His view highlights the role of models instead of theories, judgments instead of entailments, and the interdisciplinary requirements of modern science. Cartwright 1983 presented an influential critique of the character of scientific laws in explanation, exposing the partial, abstract character of such representations.

12. Parmenides and Heraclitus were pre-Socratic philosophers, who lived in the fifth century BC and whose work is primarily known from the commentaries on them by Plato and Aristotle.

CHAPTER TWO

1. Horan (1989) also identifies evolved diversity, or uniqueness as a characteristic of complex biological systems, which distinguishes them from complex physical structures. Her conclusion in light of this factor is to recommend the search for reliable generalizations by means of natural history and methods of comparison instead of the development of abstract and unrealistic theoretical models. Both her reasoning and mine were stimulated by reflecting on the work of Cartwright (1983) and Mayr (1961).

2. Even this conclusion may be suspect as a category mistake, because "coherence" and "consistency" are properties that adhere to the category of representations of the world (linguistic and otherwise) rather than to the category of things in the world.

3. Structural realism seeks to derive some of its plausibility from this line of argument. Structural realism is the view that history of science shows that while particular properties and entities described in a given theory may be superseded by a new theory, for example, the rejection of phlogiston with the acceptance of oxygen,

structural relations are preserved. Poincaré (1905) and Duhem (1906) championed this view in the early twentieth century; Worrall (1989) is a notable proponent in current discussions.

4. Dirac, one of the founders of quantum mechanics, stated in 1929 (714): "The underlying laws necessary for the mathematical theory of a large part of physics and the whole of chemistry are thus completely known, and the difficulty is only that exact applications of these laws lead to equations which are too complicated to be soluble." For a recent argument that chemistry is not even approximately reduced to physics, see Scerri 1994.

5. For a thorough discussion, see Corning 2002.

6. The introduction of the notion of downward causation marks a distinct divide between nineteenth-century emergentist thought and the use of the term in modern science, since emergent properties for the nineteenth-century philosophers had no causal connection with "lower-level" phenomena that yielded to scientific explanation. Broad thought there might be transordinal laws that connected lower levels with emergent levels, but did not discuss what we would now call downward causation. Sperry (1964) discussed what amounts to downward causation in terms of configurational forces, but the term "downward causation" is introduced by Don Campbell (1974).

7. There is an enormous literature on reduction and multiple realizability that stems from the seminal papers of Putnam (1967) and Fodor (1974). Some (e.g., Pylyshyn 1984) take the fact that instances of a "higher-level" property, like pain, can be realized in many different kinds of lower-level properties (e.g., different neurological configurations) as sufficient to argue that types of higher-level properties can never be reduced to types of lower-level properties. Others have argued that this feature is not enough to derail reductionism (see Horgan 1993; Bickle 1988). See Sober 1999 for a discussion of its merits. My argument does not address this particular thread of argumentation.

8. Bedau 1997 develops an account of weak emergence that is sensitive to the complicated way in which emergent properties are generated from their component parts. For other criticisms of the Kim-style view of emergence and reduction, see Loewer 2001; Shoemaker 2002.

9. What it means to be physical is not always clear; see Markosian 2009. For my argument all that is required is the claim that what is physical (metaphysically) is not one-to-one identical with what is represented by any given theory of physics.

10. It is important to distinguish between nonlinearity, chaos, and feedback systems. While these can all be present in a single system at the same time, they need not be. See Bak 1996; Solé and Goodwin 2001.

CHAPTER THREE

1. Sociality that includes division of labor has evolved twenty times independently in insects (Grimaldi and Engel 2004, 4).

2. See Carroll 2008 for an overview of this philosophically rich topic.

3. The most recent contender for a "biological" law that satisfies the condition of strict laws is that of allometric scaling. Though these results have been promoted as

laws, there is some dispute as to their actually satisfying the requirements. See West and Brown 2005.

4. Alleles are what we call "genes" that occur in pairs in sexually reproducing organisms at a specific site on a chromosome.

5. Laplace was a late-eighteenth-, early-nineteenth-century French mathematician who believed the universe to be completely deterministic and hypothesized an intellect who would be able to predict all future and past states from knowledge of laws of nature and the position of all particles.

6. Beatty allows that there may be degrees of contingency and that nonbiological generalizations may be contingent. My work has been to develop these ideas rigorously, having been inspired to do so in response to Beatty's important paper.

7. There are no gold spheres greater than a mile in diameter simply because there isn't that much gold on earth.

CHAPTER FOUR

1. See Strohman 2002 for an account of some of the complexity of genotype-phenotype relationships.

2. The *TP53* gene encodes the p53 protein, and it is the protein that is active in tumor suppression. Without the *TP53* gene, that protein is not produced. Yet even with a *TP53* gene, the corresponding protein can be inactivated by some interfering factor, like the human adenovirus or human papilloma virus, which binds to the p53 molecule, or via other pathways involving other genes that produce proteins that bind to and inactivate the p53 protein (Levine, Finlay, and Hinds 2004; May and May 1999).

3. See Jablonka and Lamb 1995 for details about the complexity of genes in development.

4. See Bechtel and Richardson 1993 for a detailed account of when decomposition is explanatory and when it is not.

5. DNA microarrays are tools used to analyze and measure the activity of genes. Microarrays measure gene expression by taking advantage of the process by which the component base nucleotides of DNA pair up. Only complementary bases will bind to each other, and thus a probe can detect which parts of the DNA sequence are being expressed by recording its binding behavior. See Bloom, Freyer, and Micklos 1996.

CHAPTER FIVE

1. The paradox arises in a game where a fair coin is tossed until a head appears; if the first head appears on the *n*th toss, the payoff is 2^n dollars, ducats, or pick your favorite currrency. If the coin comes up heads the first time, the prize is 2^1 = $2, and the game ends. If not, it is flipped again. If it comes up heads the second time, the prize is 2^2 = $4, and the game ends. If not, it is flipped again. And so on. The expected value of the game is infinite, yet it seems no rational person would pay to play.

2. The discussion of GM foods was published in Mitchell 2007. That paper also considers the complexity of moral pluralism in forging policy for GM foods.

CHAPTER SIX

1. Ernst Mayr (1961, 1993) distinguished between ultimate and proximate explanations to differentiate questions regarding evolutionary origins from questions about mechanisms of development. See also Beatty 1994; Mayr 1994; Ariew 2003.

2. The "for free" language comes from Stuart Kauffman (1993). He uses this locution to denote order or structure occurring spontaneously, that is, without invoking adaptive selection.

3. Recognizing multiple levels of analysis can clearly be part of an integrative, not just an isolationist, strategy. That requires not the partitioning that Sherman advocated, but a call to work simultaneously at multiple levels. See Cowan, Harter, and Kandel 2000 and Cicchetti and Dawson 2002 for a defense of a pluralist, integrative-levels approach for psychology.

REFERENCES

Allen, Myles R., Peter A. Stott, John F. B. Mitchell, Reiner Schnur, and Thomas L. Delworth. 2000. Quantifying the uncertainty in forecasts of anthropogenic climate change. *Nature* 407:617–620.

Alon, U., M. G. Surette, N. Barkai, and S. Leibler. 1999. Robustness in bacterial chemotaxis. *Nature* 397:168–171.

Alonso, J., M. C. Angermeyer, S. Bernert, R. Bruffaerts, T. S. Brugha, H. Bryson, G. Girolamo, et al. 2004. Prevalence of mental disorders in Europe: Results from the European Study of the Epidemiology of Mental Disorders (ESEMeD) project. *Acta Psychiatrica Scandinavica* 109, suppl. 420:21–27.

Alonso, W. J., and C. Schuck-Paim. 2002. Sex-ratio conflicts, kin selection, and the evolution of altruism. *Proceedings of the National Academy of Sciences of the USA* 99:6843–6847.

Alvarez, L. W. , W. Alvarez, F. Asaro, and H. V. Michel. 1980. Extraterrestrial cause for the Cretaceous-Tertiary extinction. *Science* 208:1095–1108.

Amaral, L. A. N., and J. M. Ottino. 2004. Complex networks: Augmenting the framework for the study of complex systems. *European Physical Journal* B 38:147–162.

Anderson, Peter, Claus Emmeche, Niels Finnemann, and Peder Christiansen, eds. 2000. *Downward causation: Minds, bodies, and matter*. Aarhus, Denmark: Aarhus University Press.

Andersson, Malte. 1984. The evolution of eusociality. *Annual Review of Ecology and Systematics* 15:165–189.

Annas, J. 1976. *Aristotle: Metaphysics books M and* N. Oxford: Clarendon Press.

Ariew, André. 2003. Ernst Mayr's "ultimate/proximate" distinction reconsidered and reconstructed. *Biology and Philosophy* 18:553–565.

Arrow, Kenneth. 1951. *Social choice and individual values*. New York: Wiley.

Bak, P. 1996. *How nature works: The science of self-organized criticality*. New York: Copernicus.

Bankes, Steven C. 1993. Exploratory modeling for policy analysis. *Operations Research* 41 (3): 435–449.

Bankes, Steven C., Robert J. Lempert, and Steven W. Popper. 2001. Computer-assisted reasoning. *Computing in Science and Engineering* 3:71–74.

Barkai, N., and S. Leibler. 1997. Robustness in simple biochemical networks. *Nature* 387:913–917.

Batterman, Robert. 2002. *The devil in the details: Asymptotic reasoning in explanation, reduction, and emergence*. New York: Oxford University Press.

Beatty, John. 1987. Natural selection and the null hypothesis. In *The latest on the best: Essays on evolution and optimality*, ed. J. Dupré, 53–76. Cambridge, MA: MIT Press.

———. 1994. Ernst Mayr and the proximate/ultimate distinction. *Biology and Philosophy* 9:333–356.

———. 1995. The evolutionary contingency thesis. In *Concepts, theories, and rationality in the biological sciences*, ed. Gereon Wolters and James G. Lennox, 45–81. Pittsburgh: University of Pittsburgh Press.

———. 1997. Why do biologists argue like they do? *Philosophy of Science* S64:231–242.

Bechtel, W., and R. C. Richardson. 1993. *Discovering complexity: Decomposition and localization as strategies in scientific research*. Princeton, NJ: Princeton University Press.

Beckermann, Ansgar, Hans Flohr, and Jaegwon Kim, eds. 1992. *Emergence or reduction? Essays on the prospects of nonreductive physicalism*. Berlin: Walter de Gruyter.

Bedau, Mark. 1997. Weak emergence. In *Philosophical perspectives: Mind, causation, and world*, ed. James Tomberlin, 11:375–399. Oxford: Blackwell.

———. 2003. Downward causation and autonomy in weak emergence. *Principia Revista Internacional de Epistemologica* 6:5–50.

Beilby, James K., ed. 2002. *Naturalism defeated? Essays on Plantinga's evolutionary argument against naturalism*. Ithaca, NY: Cornell University Press.

Bengel, Dietmar, Armin Heils, Susanne Petri, Marius Seemann, Katharina Glatz, Anne Andrews, Dennis L. Murphy, and K. Peter Lesch. 1997. Gene structure and 5′-flanking regulatory region of the murine serotonin transporter. *Molecular Brain Research* 44 (2): 286–292,

Benincá, E., J. Huisman, R. Heerkloss, K. D. Johnk, P. Branco, E. H. Van Nes, M. Scheffer, and S. P. Ellner. 2008. Chaos in a long-term experiment with a plankton community. *Nature* 451:822–825.

Beshers, Samuel N., and Jennifer H. Fewell. 2001. Models of division of labor in social insects. *Annual Review of Entomology* 46:413–440.

Blitz, D. 1992. *Emergent evolution: Qualitative novelty and the levels of reality*. Dordrecht, Netherlands: Kluwer Academic.

Bloom, Mark V., Greg A. Freyer, and David A. Micklos. 1996. *Laboratory DNA science: An introduction to recombinant DNA techniques and methods of genome analysis*. Menlo Park, CA: Addison Wesley.

Bodansky, Daniel. 1991. Scientific uncertainty and the precautionary principle. *Environment* 33 (7): 4–5, 43–44.

Bogen, James. 2004. Analyzing causality: The opposite of counterfactual is factual. *International Studies in the Philosophy of Science* 18:3–26.

Bonabeau, Eric, Guy Theraulaz, Jean-Louis Deneubourg, S. Aron, and Scott Camazine. 1997. Self-organization in social insects. *Trends in Ecology and Evolution* 12:188–193.

Boogerd, Fred C., Frank J. Bruggeman, Jan-Hendrick S. Hofmeyr, and Hans V. Westerhoff. 2007. *Systems biology: Philosophical foundations*. Amsterdam: Elsevier Science.

Boomsma, J. J., and L. Sundström. 1998. Patterns of paternity skew in *Formica* ants. *Behavioral Ecology and Sociobiology* 42:85–92.

Borges, Jorge Luis. 1998. *A universal history of infamy*. In *Collected fictions*, translated by Andrew Hurley. New York: Viking Penguin.

Boutilier, C., T. Dean, and S. Hanks. 1999. Decision theoretic planning: Structural assumptions and computational leverage. *Journal of AI Research* 11:1–94.

Brandon, Robert. 1997. Does biology have laws? The experimental evidence. *Philosophy of Science* 64:S444–S457.

———. 2006. The principle of drift: Biology's first law. *Journal of Philosophy* 102 (7): 319–335.

Breed, Michael, and Robert E. Page, Jr. 1989. *The genetics of social evolution*. Boulder, CO: Westview Press.

Broad, Charles Dunbar. 1925. *The mind and its place in nature*. London: Routledge and Kegan Paul.

Buck, Louise E., Charles C. Geisler, John Schelhas, and Eva Wollenberg, eds. 2001. *Biological diversity: Balancing interests through adaptive collaborative management*. London: Taylor and Francis / CRC Press.

Buss, Leo. 1987. *The evolution of individuality*. Princeton, NJ: Princeton University Press.

Butts, Robert E., ed. 1989. *William Whewell: Theory of scientific method*. Indianapolis, IN: Hackett Publishing Co.

Byrne, David. 1998. *Complexity theory and the social sciences: An introduction*. New York: Routledge.

Camazine, Scott, Jean-Louis Deneubourg, Nigel R. Franks, James Sneyd, Guy Theraulaz, and Eric Bonabeau. 2001. *Self-organization in biological systems*. Princeton, NJ: Princeton University Press.

Campbell, Donald T. 1974. "Downward causation" in hierarchically organized biological systems. In *Studies in the philosophy of biology*, ed. F. J. Ayala and T. Dobzhansky, 179–186. New York: Macmillan Press.

Carnap, Rudolf. 1950. *Logical foundations of probability*. Chicago: University of Chicago Press.

Carroll, John W. 2008. Laws of nature. In *The Stanford Encyclopedia of Philosophy*, fall 2008 edition, ed. Edward N. Zalta. http://plato.stanford.edu/archives/fall2008/entries/laws-of-nature/.

Cartwright, Nancy D. 1983. *How the laws of physics lie*. Oxford: Oxford University Press.

———. 1989. *Nature's capacities and their measurement.* Oxford: Oxford University Press.

———. 1994. Fundamentalism vs. the patchwork of laws. *Proceedings of the Aristotelian Society* 94:279–292.

———. 2000. *The dappled world: A study of the boundaries of science.* Cambridge: Cambridge University Press.

———. 2002. Against modularity, the causal Markov condition and any link between the two: Comments on Hausman and Woodward. *British Journal for the Philosophy of Science* 53 (3): 411–453.

Cartwright, Nancy D., Jordi Cat, Lola Fleck, and Thomas E. Uebel. 1995. *Philosophy between science and politics.* Ideas in Context Series. Cambridge: Cambridge University Press.

Caspi, Avshalom, Karen Sugden, Terrie E. Moffitt, Alan Taylor, Ian W. Craig, Hona-Lee Harrington, Joseph McClay, et al. 2003. Influence of life stress on depression: Moderation by a polymorphism in the 5-*HTT* gene. *Science* 301:386–389.

Chang, Hasok. 2004. *Inventing temperature: Measurement and scientific progress.* New York: Oxford University Press.

Cicchetti, Dante, and Geraldine Dawson. 2002. Editorial: Multiple levels of analysis. *Development and Psychopathology* 14:417–420.

Ciliberti, Stefano, Olivier C. Martin, and Andreas Wagner. 2007. Robustness can evolve gradually in complex regulatory gene networks with varying topology. *PLoS Computational Biology* 3 (2): e15.

Clark, James S., Steven R. Carpenter, Mary Barber, Scott Collins, Andy Dobson, Jonathan A. Foley, David M. Lodge, et al. 2001. Ecological forecasts: An emerging imperative. *Science* 293:657–660.

Clayton, Philip, and Paul Davies, eds. 2006. *The re-emergence of emergence.* Oxford: Oxford University Press.

Collins, Harry M., ed. 1982. *The sociology of scientific knowledge: A sourcebook.* Bath, UK: Bath University Press.

———. 1985. *Changing order: Replication and induction in scientific practice.* Beverly Hills, CA: Sage.

Corning, Peter. 2002. The re-emergence of "emergence": A venerable concept in search of a theory. *Complexity* 7 (6): 18–30.

Costanza, Roberto, and Laura Cornwell. 1992. The 4P approach to dealing with scientific uncertainty. *Environment* 34 (9): 12–20.

Couzin, Iain D. 2007. Collective minds. *Nature* 445:4.

Couzin, Iain D., and Jens Krause. 2003. Self-organization and collective behavior in vertebrates. *Advances in the Study of Behavior* 33:1–75.

Cowan, W. M., D. H. Harter, and E. R. Kandel. 2000. The emergence of modern neuroscience: Some implications for neurology and psychiatry. *Annual Review of Neuroscience* 23:323–391.

Cranor, Carl. 2001. Learning from the law to address uncertainty in the precautionary principle. *Science and Engineering Ethics* 7:313–326.

Craver, Carl F. 2007. *Explaining the brain: Mechanisms and the mosaic unity of neuroscience.* Oxford: Oxford University Press.

Cushing, J. T., and E. McMullin, eds. 1989. *Philosophical consequences of quantum theory.* South Bend, IN: University of Notre Dame Press.

Darwin, Charles. 1839. *The voyage of the Beagle.* Reprint. New York: Modern Library, 2001.

———. 1859. *On the origin of species by means of natural selection; or, The preservation of favoured races in the struggle for life.* London: John Murray.

Dawkins, Richard. 1996. *Climbing Mount Improbable.* New York: Norton and Co.

Delehanty, Megan. 2005. Emergent properties and the context objection to reduction. *Biology and Philosophy* 20:715–734.

Delgado, P. L. 2000. Depression: The case for a monoamine deficiency. *Journal of Clinical Psychiatry* 61, suppl. 6:7–11.

Dennett, Daniel. 1995. *Darwin's dangerous idea: Evolution and the meanings of life.* New York: Simon and Schuster.

Diagnostic and statistical manual of mental disorders (DSM-IV). 2000. 4th edition. Washington, DC: American Psychiatric Publishing.

Diamond, E. 1999. Genetically modified organisms and monitoring. *Journal of Environmental Monitoring* 1:108N–110N.

Dirac, P. A. M. 1929. Quantum mechanics of many-electron systems. *Proceedings of the Royal Society of London* A 123:714–733.

Duhem, P. 1906. *The aim and structure of physical theory.* Reprint. Princeton, NJ: Princeton University Press, 1991.

Dupré, John. 1993. *The disorder of things: Metaphysical foundations of the disunity of science.* Cambridge, MA: Harvard University Press.

———. 2002. Is "natural kind" a natural kind term? *Monist* 85:29–49.

Earman, John, John Roberts, and Sheldon Smith. 2002. Ceteris paribus lost. *Erkenntnis* 57:281–301.

Eaves, L. J. 2006. Genotype × environment interaction in psychopathology: Fact or artifact? *Twin Research and Human Genetics* 9:1–8.

Edelman, Gerald M., and Joseph A. Gally. 2001. Degeneracy and complexity in biological systems. *Proceedings of the National Academy of Sciences of the USA* 98:13763–13768.

Emmeche, Claus. 1997. Aspects of complexity in life and science. *Philosophica* 59:41–68.

Esogbue, Augustine O., and Barry Randall Marks. 1974. Non-serial dynamic programming: A survey. *Operational Research Quarterly* 25 (2): 253–265.

Eve, Raymond A., Sara Horsfall, and Mary E. Lee, eds. 1997. *Chaos, complexity and sociology: Myths, models and theories.* Thousand Oaks, CA: Sage.

Falk, Rafael. 1996. What is a gene? *Studies in the History and Philosophy of Science* 17:133–173.

Feder, Toni. 2007. Statistical physics is for the birds. *Physics Today* 60:28–30.

Fewell, Jennifer H., and Robert E. Page Jr. 1999. The emergence of division of labour in forced associations of normally solitary ant queens. *Evolutionary Ecology Research* 1:537–548.

Feyerabend, Paul K. 1975. *Against method: Outline of an anarchistic theory of knowledge.* London: New Left Books.

———. 1981. *Philosophical papers*. Cambridge: Cambridge University Press.

Fodor, Jerry. 1974. Special sciences; or, The disunity of science as a working hypothesis. *Synthese* 28:97–115.

Foitzik, S., and J. M. Herbers. 2001. Colony structure of a slavemaking ant: part 1, Intra-colony relatedness, worker reproduction and polydomy. *Evolution* 55:307–315.

Fontana, Walter, and Leo W. Buss. 1994. What would be conserved if "the tape were played twice"? *Proceedings of the National Academy of Sciences of the USA* 91:757–761.

Forster, Lucy, Peter Forster, Sabine Lutz-Bonengel, Horst Willkomm, and Bernd Brinkmann. 2002. Natural radioactivity and human mitochondrial DNA mutations. *Proceedings of the National Academy of Sciences of the USA* 99:13950–13954.

Foster, K., P. Vecchia, and M. Repacholi. 2000. Science and the precautionary principle. *Science* 288:979–981.

Fox Keller, Evelyn. 2002. *Making sense of life: Explaining biological development with models, metaphors, and machines*. Cambridge, MA: Harvard University Press.

Friedman, Michael. 1974. Explanation and scientific understanding. *Journal of Philosophy* 71:5–19.

Galison, Peter, and David J. Stump, eds. 1996. *The disunity of science: Boundaries, contexts, and power*. Stanford, CA: Stanford University Press.

Gannett, Lisa. 1999. What's in a cause? The pragmatic dimensions of genetic explanations. *Biology and Philosophy* 14:349–374.

Gerstein, Mark B., Can Bruce, Joel S. Rozowsky, Deyou Zheng, Jiang Du, Jan O. Korbel, Olof Emanuelsson, Zhengdong D. Zhang, Sherman Weissman, and Michael Snyder. 2007. What is a gene, post-ENCODE? History and updated definition. *Genome Research* 17:669–681.

Giarratani, F., G. Gruver, and R. Jackson. 2007. Clusters, agglomeration, and economic development potential: Empirical evidence based on the advent of slab casting by U.S. steel minimills. *Economic Development Quarterly* 21:148–164.

Giere, Ronald. 1999. *Science without laws: Science and its conceptual foundations*. Chicago: University of Chicago Press.

Gigerenzer, Gerd. 1996. On narrow norms and vague heuristics: A reply to Kahneman and Tversky (1996). *Psychological Review* 103:592–596.

Goldbeter, A. 1997. *Biochemical oscillations and cellular rhythms: The molecular bases of periodic and chaotic behaviour*. Cambridge: Cambridge University Press.

Goldbeter, A., D. Gonze, G. Houart, J. C. Leloup, J. Halloy, and G. Dupont. 2001. From simple to complex oscillatory behaviorin metabolic and genetic control network. *Chaos* 11:247–260.

Goodman, Nelson. 1947. The problem of counterfactual conditionals. *Journal of Philosophy* 44 (5): 113–128.

Gordon, Deborah M. 1989. Dynamics of task switching in harvester ants. *Animal Behaviour* 38:194–204.

Gould, Stephen Jay. 1990. *Wonderful life: Burgess Shale and the nature of history*. New York: W. W. Norton.

———. 1997. Evolution: The pleasures of pluralism. *New York Review of Books*, June 26, 47–52.

Gower, Barry. 1997. *Scientific method: An historical and philosophical introduction*. New York: Routledge.

Grant, Peter, and Rosemary Grant. 1986. *Ecology and evolution of Darwin's finches*. Princeton, NJ: Princeton University Press.

Greenspan, Ralph J. 2001. The flexible genome. *Nature Reviews Genetics* 2:383–387.

Griffiths, Paul E., and Karola Stotz. 2007. Gene. In *Cambridge companion to philosophy of biology*, ed. Michael Ruse and David Hull, 85–102. Cambridge: Cambridge University Press.

Grimaldi, David, and Michael S. Engel. 2004. *Evolution of the insects*. Cambridge: Cambridge University Press.

Hacking, Ian. 1993. Working in a new world: The taxonomic solution. In *World changes: Thomas Kuhn and the nature of science*, ed. Paul Horwich, 275–310. Cambridge, MA: MIT Press.

Haddjeri, Nasser P. B., and C. de Montigny. 1998. Long-term antidepressant treatments result in a tonic activation of forebrain 5-HT1A receptors. *Journal of Neuroscience* 18:10150–10156.

Hamilton, William D. 1964. The genetical evolution of social behaviour I and II. *Journal of Theoretical Biology* 7:1–16, 17–32.

Hanson, Norwood Russell. 1958. *Patterns of discovery: An inquiry into the conceptual foundations of science*. Cambridge: Cambridge University Press.

Hariri, Ahmad R., Emily M. Drabant, Karen E. Munoz, Bhaskar S. Kolachana, Venkata S. Mattay, Michael F. Egan, and Daniel R. Weinberger. 2005. A susceptibility gene for affective disorders and the response of the human amygdala. *Archives of General Psychiatry* 62:146–152.

Hausman, Daniel, and James Woodward. 1999. Independence, invariance and the causal Markov condition. *British Journal of the Philosophy of Science* 50:521-583.

Hempel, Carl G., and Paul Oppenheim. 1948. Studies in the logic of explanation. *Philosophy of Science* 15:135–175. Reprinted with a postscript in *Aspects of Scientific Explanation and other essays*, ed. Carl G. Hempel, 245–295. New York: Free Press, 1965.

Herschel, John. 1830. *Preliminary discourse on the study of natural philosophy*. London: Thoemmes Press.

Holliday, C. M., R. C. Ridgely, A. M. Balanoff, and L. M. Witmer. 2006. Cephalic vascular anatomy in flamingos (*Phoenicopterus ruber*) based on novel vascular injection and computed tomographic imaging analyses. *Anatomical Record* 288A:1031–1041.

Horan, Barbara L. 1989. Theoretical models, biological complexity and the semantic view of theories. In *PSA: Proceedings of the Biennial Meeting of the Philosophy of Science Association 1988*, vol. 2, *Symposia and invited papers*, 265–277. Chicago: University of Chicago Press.

Horgan, Terence. 1993. Nonreductive materialism and the explanatory autonomy of psychology. In *Naturalism: A critical appraisal*, ed. Steven Wagner and Richard Warner, 295–320. Notre Dame, IN: University of Notre Dame Press.

Humphreys, Paul. 1997. How properties emerge. *Philosophy of Science* 64:1–17.

Ihle, James N. 2000. The challenges of translating knockout phenotypes into gene function. *Cell* 102:131–134.

International Panel on Climate Change (IPCC). 1995. *Climate change 1995: The science of climate change; Contribution of Working Group I to the second assessment report of the Intergovernmental Panel on Climate Change*, ed. J. T. Houghton, L. G. Meira Filho, B. A. Callander, N. Harris, A. Kattenberg, and K. Maskell. Cambridge: Cambridge University Press.

———. 2001. *Climate change 2001: Synthesis report; Contribution of Working Groups I, II, and III to the third assessment report of the Intergovernmental Panel on Climate Change*, [ed. R. T. Watson and the Core Writing Team]. Cambridge: Cambridge University Press.

———. 2007. *Climate change 2007: Impacts, adaptation and vulnerability; Contribution of Working Group II to the fourth assessment report of the Intergovernmental Panel on Climate Change*, ed. M. L. Parry, O. F. Canziani, J. P. Palutikof, P. J. van der Linden, and C. E. Hanson. Cambridge: Cambridge University Press.

Ishii, Nobuyoshi, Kenji Nakahigashi, Tomoya Baba, Martin Robert, Tomoyoshi Soga, Akio Kanai, Takashi Hirasawa, et al. 2007. Multiple high-throughput analyses monitor the response of *E. coli* to perturbations. *Science* 316:593–597.

Jablonka, Eva, and Marion Lamb. 1995. *Epigenetic inheritance and evolution: The Lamarckian dimension*. Oxford: Oxford University Press.

Jacob, François, and Jacques Monod. 1961. Genetic regulatory mechanisms in the synthesis of proteins. *Journal of Molecular Biology* 3:318–356.

James, William. 1890. *The principles of psychology*. New York: Henry Holt and Co.

Jansen, Ritsert C. 2003. Opinion: Studying complex biological systems using multifactorial perturbation. *Nature Reviews Genetics* 4:145–151.

Jasny, Michael, Joel Reynolds, and Ann Notthoff. 1997. *Leap of faith: Southern California's experiment in natural community conservation planning*. New York: Natural Resources Defense Council.

Jeanson, Raphaël, and Jennifer H. Fewell. 2008. Influence of the social context on division of labor in ant foundress associations. *Behavioral Ecology* 19 (3): 567–574.

Jeffrey, Richard. 1990. *The logic of decision*. Chicago: University of Chicago Press.

Kahneman, Daniel, and Amos Tversky. 1973. On the psychology of prediction. *Psychological Review* 80:237–251.

Kaiser, Matthias. 1997. The precautionary principle and its implications for science. *Foundations of Science* 2:201–205.

Kauffman, Stuart A. 1984. Emergent properties in random complex automata. *Physica D: Nonlinear Phenomenon* 10:145–156.

———. 1993. *The origins of order*. Oxford: Oxford University Press.

———. 1995. *At home in the universe: The search for laws of self-organization and complexity*. New York: Oxford University Press.

———. 2008. *Reinventing the sacred: A new view of science, reason, and religion*. New York: Basic Books.

Kellert, Stephen H., Helen E. Longino, and C. Kenneth Waters, eds. 2006. *Scientific pluralism*. Minnesota Studies in the Philosophy of Science. Minneapolis: University of Minnesota Press.

Kendler, Kenneth S. 2005a. "A gene for...": The nature of gene action in psychiatric disorders. *American Journal of Psychiatry* 162:1243–1252.

———. 2005b. Toward a philosophical structure for psychiatry. *American Journal of Psychiatry* 162:433–440.

Kendler, Kenneth S., C. O. Gardner, and Carol A. Prescott. 2002. Toward a comprehensive developmental model for major depression in women. *American Journal of Psychiatry* 159:1133–1145.

———. 2006. Toward a comprehensive developmental model for major depression in men. *American Journal of Psychiatry* 163:115–124.

Kendler, Kenneth S., Jonathan W. Kuhn, Jen Vittum, Carol A. Prescott, and Brien Riley. 2005. The interaction of stressful life events and a serotonin transporter polymorphism in the prediction of episodes of major depression A replication. *Archives of General Psychiatry* 62:529–535.

Kendler, Kenneth S., and Josef Parnas, eds. 2008. *Philosophical issues in psychiatry: Explanation, phenomenology and nosology*. Baltimore: Johns Hopkins University Press.

Kenyon, Jane. 1996. Having it out with melancholy. In *Otherwise*, 189–193. St. Paul, MN: Graywolf Press.

Kessin, R. H. 2001. *Dictyostelium: Evolution, cell biology, and the development of multicellularity*. Cambridge: Cambridge University Press.

Kessler, Ronald C., Patricia Berglund, Olga Demler, Robert Jin, Doreen Koretz, Kathleen R. Merikangas, A. John Rush, Ellen E. Walters, and Philip S. Wang. 2003. The epidemiology of major depressive disorder: Results from the National Comorbidity Survey Replication (NCS-R). *Journal of the American Medical Association* 289:3095–3105.

Keynes, John M. 1921. *A treatise on probability*. London: Macmillan.

Kim, Jaegwon. 1999. Making sense of emergence. *Philosophical Studies* 95:3–36.

Kirk, G. S., J. E. Raven, and M. Schofield. 1983. *The Presocratic philosophers*. 2nd edition. Cambridge: Cambridge University Press.

Kitano, Hiroaki. 2002. Systems biology: A brief overview. *Science* 295:1662–1664.

———. 2004. Biological robustness. *Nature Reviews Genetics* 5:826–837.

Kitcher, Philip. 1981. Explanatory unification. *Philosophy of Science* 48:507–531.

Korb, Kevin B., and Ann E. Nicholson. 2003. *Bayesian artificial intelligence*. London: Taylor and Francis / CRC Press.

Kragh, Helge. 1999. *Quantum generations: A history of physics in the twentieth century*. Princeton, NJ: Princeton University Press.

Krieger, Nancy. 1994. Epidemiology and the web of causation: Has anyone seen the spider? *Social Science and Medicine* 39:887–903.

Krüger, Lorenz, Lorraine Daston, and Michael Heidelberger. 1990. *The probabilistic revolution*. Vol. 1, *Ideas in history*. Cambridge, MA: Bradford Books.

Kuhn, Thomas. 1962. *The structure of scientific revolutions*. Chicago: University of Chicago Press.

Lacey, Hugh. 1999. *Is science value free? Values and scientific understanding*. London: Routledge.

———. 2002. Assessing the value of transgenic crops. *Ethics in Science and Technology* 8:497–511.

Lakatos, Imre. 1970. Falsification and the methodology of scientific research programmes. In *Criticism and the growth of knowledge*, ed. Imre Lakatos and Alan Musgrave, 91–196. Cambridge: Cambridge University Press.

Lakatos, Imre, and Alan Musgrave, eds. 1970. *Criticism and the growth of knowledge*. Cambridge: Cambridge University Press.

Lange, Marc. 2000. *Natural laws in scientific practice*. Oxford: Oxford University Press.

Latour, Bruno. 1988. *Science in action*. Cambridge, MA: Harvard University Press.

Laughlin, Robert. 2005. *A different universe: Reinventing physics from the bottom up*. New York: Basic Books.

Lempert, Robert J. 2002. A new decision science for complex systems. *Proceedings of the National Academy of Science of the USA* 99, suppl. 3: 7309–7313.

Lempert, Robert J., Steven W. Popper, and Steven C. Bankes. 2002. Confronting surprise. *Social Science Computer Review* 20:420–440.

Levine, Arnold J., Cathy A. Finlay, and Philip W. Hinds. 2004. *P53* is a tumor suppressor gene. *Cell* 116:S67–S70.

Lewin, Roger. 1992. *Complexity: Life at the edge of chaos*. Chicago: University of Chicago Press.

Lewontin, Richard C. 2003. Science and simplicity. *New York Review of Books*, May 1.

Li, F. P., and J. F. Fraumeni Jr. 1969. Soft-tissue sarcomas, breast cancer and other neoplasms: A familial syndrome? *Annals of Internal Medicine* 71:747–752.

Loewer, Barry. 2001. Review of J. Kim, *Mind in a physical world*. *Journal of Philosophy* 98:315–324.

Longino, Helen. 2002. *The fate of knowledge*. Princeton, NJ: Princeton University Press.

Lorant, Vincent, Christophe Croux, Scott Weich, Denise Deliège, Johan Mackenbach, and Marc Ansseau. 2007. Depression and socio-economic risk factors: 7-year longitudinal population study. *British Journal of Psychiatry* 190:293–298.

Lorenz, E. N. 1996. *The essences of chaos*. Seattle: University of Washington Press.

Mackie, J. L. 1965. Causes and conditions. *American Philosophical Quarterly* 2 (4): 245–264.

MacMahon, B., T. F. Pugh, and J. Ipsen. 1960. *Epidemiologic methods*. Boston: Little Brown and Co.

Malka, O., Shiri Shnieor, Abraham Heftz, and Tamar Katzav-Gozansky. 2007. Reversible royalty in worker honeybees (*Apis mellifera*) under the queen influence. *Behavioral Ecology and Sociobiology* 61:465–473.

Mangel, Marc, Lee M. Talbot, Gary K. Meffe, M. Tundi Agardy, Dayton L. Alverson, Jay Barlow, Daniel B. Botkin, et al. 1996. Principles for conservation of wild living resources. *Ecological Applications* 6:338–362.

Mantzavinos, Chrysostomos. 2009. *Philosophy of the social sciences: Philosophical theory and scientific practice*. Cambridge: Cambridge University Press.

Markosian, Ned. 2009. *Physical object: A companion to metaphysics*. 2nd edition. Ed. Ernest Sosa, Jaegwon Kim, and Gary Rosenkrantz. Oxford: Blackwell Publishing.

Marra, Michele C., Philip G. Pardey, and Julian M. Alston. 2003. The payoffs to transgenic field crops: An assessment of the evidence. *AgBioForum* 5 (2): 43–50.

May, Pierre, and Evelyne May. 1999. Twenty years of p53 research: Structural and functional aspects of the p53 protein. *Oncogene* 18:7621–7636.

May, R. M., and G. Oster. 1976. Bifurcations and dynamic complexity in simple ecological models. *American Naturalist* 110:573.

Maynard Smith, J., and E. Szathmáry. 1993. Proximate and ultimate causations. *Biology and Philosophy* 8:93–94.

———. 1997. *The major transitions in evolution.* New York: Oxford University Press

Mayr, Ernst. 1961. Cause and effect in biology. *Science* 134:1501–1506.

———. 1993. Proximate and ultimate causation. *Biology and Philosophy* 8:95–98.

———. 1994. Response to John Beatty. *Biology and Philosophy* 9:359–371.

McLaughlin, Brian. 1992. The rise and fall of British emergentism. In *Emergence or reduction?* ed. Ansgar Beckermann, Hans Flohr, and Jaegwon Kim, 49–93. Berlin: Walter de Gruyter.

McShea, Daniel W. 2005. A universal generative tendency toward increased organismal complexity. In *Variation: A central concept in biology*, ed. B. Hallgrimsson and B. Hall, 435–453. San Diego, CA: Academic Press.

Michener, Charles D., and Denis J. Brothers. 1974. Were workers of eusocial Hymenoptera initially altruistic or oppressed? *Proceeding of the National Academy of Science of the USA* 71:671–674.

Mill, John Stuart. 1843. *A system of logic.* London: Longmans, Green and Co.

Mitchell, Sandra D. 1993. Dispositions or etiologies: A comment on Bigelow and Pargetter. *Journal of Philosophy* 90 (5): 249–259.

———. 1997. Pragmatic laws. In *PSA: Proceedings of the Biennial Meeting of the Philosophy of Science Association* 1996, vol. 2, *Symposia and invited papers*, 468–479.

———. 2000. Dimensions of scientific law. *Philosophy of Science* 67:242–265.

———. 2002a. Ceteris paribus: An inadequate representation for biological contingency. *Erkenntnis* 57:329–350.

———. 2002b. Contingent generalizations: Lessons from biology. In *Akteure, Mechanismen, Modelle: Zur Theoriefähigkeit makro-sozialer Analysen*, ed. R. Mayntz, 179–195. Frankfurt: Campus Verlag.

———. 2002c. Integrative pluralism. *Biology and Philosophy* 17:55–70.

———. 2003. *Biological complexity and integrative pluralism.* Cambridge: Cambridge University Press.

———. 2007. The import of uncertainty. *Pluralist* 2:58–71.

———. 2008a. Comment: Taming causal complexity. In *Philosophical issues in psychiatry: Explanation, phenomenology and nosology*, ed. Kenneth S. Kendler and Josef Parnas, 125–131. Baltimore: Johns Hopkins University Press.

———. 2008b. Explaining complex behavior. In *Philosophical issues in psychiatry: Explanation, phenomenology and nosology*, ed. Kenneth S. Kendler and Josef Parnas, 19–47. Baltimore: Johns Hopkins University Press.

———. 2008c. Exporting causal knowledge in evolutionary and developmental biology. *Philosophy of Science* 75:697–706.

Moffitt, Terrie E., Avshalom Caspi, and Michael Rutter. 2005. A research strategy for investigating interactions between measured genes and measured environments. *Archives of General Psychiatry* 62:769–775.

Morgan, M. G., and M. Henrion. 1990. *Uncertainty: A guide to dealing with uncertainty in quantitative risk and policy analysis*. Cambridge: Cambridge University Press.

Moser, Paul K., and J. D. Trout, ed. 1995. *Contemporary materialism: A reader*. New York: Routledge.

Müller, U. 1999. Ten years of gene targeting: Targeted mouse mutants, from vector design to phenotype analysis. *Mechanisms of Development* 82:3–21.

Murphy, Dominick. 2008. Levels of explanation in psychiatry. In *Philosophical issues in psychiatry: Explanation, phenomenology and nosology*, ed. Kenneth S. Kendler and Josef Parnas, 99–124. Baltimore: Johns Hopkins University Press.

Murray, Carol, and David Marmorek. 2003. Adaptive management: A science-based approach to managing ecosystems in the face of uncertainty. In *Making ecosystem based management work*, proceedings of the Fifth International Conference on Science and Management of Protected Areas, Victoria, British Columbia, May 11–16. Science and Management of Protected Areas Association. http://www.sampaa.org/PDF/ch8/8.9.pdf.

Myers, Norman. 1979. *The sinking ark: A new look at the problem of disappearing species*. New York: Pergamon Press.

———. 1995. Environmental unknowns. *Science* 269:358–360.

Nagel, Ernest. 1961. *The structure of science: Problems in the logic of scientific explanation*. New York: Harcourt, Brace and World.

National Academy. 2000. *Genetically modified pest-protected plants: Science and regulation*. Washington, DC: National Academy Press.

Nelson, Randy J. 1997. The use of genetic "knockout" mice in behavioral endocrinology research. *Hormones and Behavior* 31 (3): 188–196.

Nicolis, Gregoire, and Ilya Prigogine. 1989. *Exploring complexity: An introduction*. New York: W. H. Freeman.

Nowak, Martin A., Martin C. Boerlijst, Jonathan Cooke, and J. Maynard Smith. 1997. Evolution of genetic redundancy. *Nature* 388:167–171.

O'Connor, Timothy. 1994. Emergent properties. *American Philosophical Quarterly* 31:91–104.

O'Connor, Timothy, and Hong Yu Wong. 2005. The metaphysics of emergence. *Noûs* 39:658–678.

Oglethorpe, Judy A. E., ed. 2002. *Adaptive management: From theory to practice*. Gland, Switzerland: International Union for Conservation of Nature.

Oreskes, Naomi. 2004. Beyond the ivory tower: The scientific consensus on climate change. *Science* 306:1686.

O'Riordan, Tim, and James Cameron, eds. 1994. *Interpreting the precautionary principle*. London: Earthscan.

Oster, G. H., and E. O. Wilson. 1979. *Caste and ecology in the social insects*. Princeton, NJ: Princeton University Press.

Ottman, R., and D. C. Rao. 1990. An epidemiologic approach to gene-environment interaction. *Genetic Epidemiology* 7 (3): 177–185.

Oyama, Susan, Paul Griffiths, and Russell D. Gray, eds. 2001. *Cycles of contingency: Developmental systems and evolution*. Cambridge, MA: MIT Press.

Page, Robert E. Jr., and Robert A. Metcalf. 1982. Multiple mating, sperm utilization, and social evolution. *American Naturalist* 119:263–281.

Page, Robert E. Jr., and Sandra D. Mitchell. 1991. Self organization and adaptation in insect societies. In In *PSA: Proceedings of the Biennial Meeting of the Philosophy of Science Association 1996*, vol. 2, *Symposia and invited papers*, 289–298. Chicago: University of Chicago Press.

———. 1998. Self organization and the evolution of division of labor. *Apidologie* 29:101–120.

Patel, Dinshaw J., L. Kampa, R. G. Shulman, T. Yamane, and B. J. Wyluda. 1970. Proton nuclear magnetic resonance studies of myoglobin in H_2O. *Proceedings of the National Academy of Sciences of the USA* 67:1109–1115.

Pauling, Linus, Harvey A. Itano, S. J. Singer, and Ibert C. Wells. 1949. Sickle-cell anemia, a molecular disease. *Science* 110:543–548.

Perini, Laura. 2005. Explanation in two dimensions: Diagrams and biological explanation. *Biology and Philosophy* 20:257–269.

Peterson, Martin. 2008. *Non-Bayesian decision theory*. Berlin: Springer.

Poincaré, H. 1905. *Science and hypothesis*. Reprint. New York: Dover, 1952.

Popper, Steven W., Robert J. Lempert, and Steven C. Bankes. 2005. Shaping the future. *Scientific American* 292 (4): 66–71.

Pratt, John W., Howard Raiffa, and Robert Schlaifer. 1995. *Introduction to statistical decision theory*. Cambridge, MA: MIT Press.

Prigogine, Ilya. 1997. *The end of certainty*. New York: Free Press.

Pryce, Christopher R., Daniela Rüedi-Bettschen, Andrea C. Dettling, and Joram Feldon. 2002. Early life stress: Long-term physiological impact in rodents and primates. *News Physiological Sciences* 17:150–155.

Putnam, Hilary. 1967. Psychological predicates. In *Art, mind, and religion*, ed. W. H. Capitan and D. D. Merrill, 37–48. Pittsburgh: University of Pittsburgh Press.

Pylyshyn, Zenon. 1984. *Computation and cognition*. Cambridge, MA: MIT Press.

Raff, Rudolf A. 2000. Evo-devo: The evolution of a new discipline. *Nature Reviews Genetics* 1:74–79.

Ramachandran, V. S. 1993. Behavioral and magnetoencephalographic correlates of plasticity in the adult human brain. *Proceedings of the National Academy of Sciences of the USA* 90:10413–10420.

———. 2007. When brain maps college. In *Neuroscience: Exploring the brain*, 3rd edition, ed. M. F. Bear, B. W. Connors, and M. A. Paradiso, 406. Philadelphia: Lippencott Williams and Wilkins.

Rapoport, Anatol. 1989. *Decision theory and decision behaviour*. Berlin: Springer.

Revkin, Andrew C. 2007. A new middle stance emerges in debate over climate. *New York Times*, January 1.

Risch, Neil, Richard Herrell, Thomas Lehner, Kung-Yee Liang, Lindon Eaves, Josephine Hoh, Andrea Griem, Maria Kovacs, Jurg Ott, Kathleen Ries Merikangas. 2009. Interaction between the serotonin transporter gene (5-*HTTLPR*), stressful life events, and risk of depression: A meta-analysis. *Journal of the American Medical Association* 301(23):2462–71.

Robinson, Gene E. 1992. Regulation of division of labor in insect societies. *Annual Review of Entomology* 37:637–665.

Robinson, Gene E., C. M. Grozinger, and C. W. Whitfield. 2005. Sociogenomics: Social life in molecular terms. *Nature Reviews Genetics* 6:257–270.

Rosen, Jonathan. 2007. Flight patterns. *New York Times Magazine*, April 22.

Rosen, Robert. 1985. Organisms as causal systems which are not mechanisms: An essay into the nature of complexity. In *Theoretical biology and complexity: Three essays on the natural philosophy of complex systems*, 165–203. London: Academic Press.

Rouvray, Dennis H., and R. Bruce King, eds. 2004. *The periodic table: Into the 21st century*. Baldock, UK: Research Studies Press.

Rueger, Alexander. 2000. Physical emergence, diachronic and synchronic. *Synthese* 124:297–322.

Russell, Bertrand. 1927. *The analysis of matter*. London: Allen and Unwin.

Sandler, Todd. 1993. *Collective action: Theory and applications*. Ann Arbor: University of Michigan Press.

Sarewitz, Daniel, and Roger A. Pielke Jr. 2000. Prediction in science and policy. In *Prediction: Science, decision making, and the future of nature*, ed. Daniel Sarewitz, Roger A. Pielke Jr., and Radford Byerly, 11–22. Washington, DC: Island Press.

Sarkar, Sahotra. 1998. *Genetics and reductionism*. Cambridge: Cambridge University Press.

Sauer, Uwe, Matthias Heinemann, and Nicola Zamboni. 2007. Getting closer to the whole picture. *Science* 316:550–551.

Savage, L. 1954. *The foundations of statistics*. New York: Wiley.

Scerri, Eric R. 1994. Has chemistry been at least approximately reduced to quantum mechanics? In *PSA: Proceedings of the Biennial Meeting of the Philosophy of Science Association* 1994, vol. 1, *Contributed papers*, 160–170. Chicago: University of Chicago Press.

Schaffner, Kenneth F. 1993. *Discovery and explanation in biology and medicine*. Chicago: University of Chicago Press.

———. 2002. Reductionism, complexity and molecular medicine: Genetic chips and the "globalization" of the genome. In *Promises and limits of reductionism in the biomedical sciences*, ed. M. Regenmortel and D. Hull, 323–351. London: John Wylie.

———. Forthcoming. *Behaving: What's genetic and what's not, and why should we care?* Oxford: Oxford University Press.

Schatzberg, Alan F. 2002. Major depression. *American Journal of Psychiatry* 159:1077–1079.

Schouten, M. K. D., and H. L. de Jong. 2001. Pluralism and heuristic identification: Some explorations in behavioral genetics. *Theory and Psychology* 11:796–807.

Schwartz, Jeffrey H. 2004. Trying to make chimpanzees into humans. *History and Philosophy of the Life Sciences* 26:271–277.

Schwartz, Jeffrey H., Robert B. Eckhardt, Adrian E. Friday, Phillip D. Gingerich, J. W. Osborn, Pierre-François Puech, R. Stanyon, S. M. Borgognini Tarli, and Jan Wind. 1984. Hominoid evolution: A review and a reassessment [and comments and replies]. *Current Anthropology* 25:655–672.

Searle, John R. 1999. The future of philosophy. *Philosophical Transactions of the Royal Society London* B54:2069–2080.

Seeley, Thomas D. 1989. Social foraging in honey bees: How nectar foragers assess their colony's nutritional status. *Behavioral Ecology and Sociobiology* 24:181–199.

Seeley, Thomas D., and C. A. Tovey. 1994. Why search time to find a food-storer bee accurately indicates the relative rates of nectar collecting and nectar processing in honey bee colonies. *Animal Behavior* 47:311–316.

Selous, Edmund. 1905. *Bird life glimpses*. London: George Allen.

Sherman, Paul. 1988. The levels of analysis. *Animal Behavior* 36:616–619.

Shilling, F. 1997. Do habitat conservation plans protect endangered species? *Science* 276:1662–1663.

Shoemaker, Sydney. 2002. Kim on emergence. *Philosophical Studies* 108:53–63.

Shrader-Frechette, Kristin. 1991. *Risk and rationality*. Berkeley: University of California Press.

Silberstein, Michael. 2002. Reduction, emergence and explanation. In *The Blackwell guide to the philosophy of science*, ed. Peter K. Machamer and Michael Silberstein, 80–107. Oxford: Blackwell.

Silberstein, Michael, and John McGeever. 1999. The search for ontological emergence. *Philosophical Quarterly* 49:182–200.

Simon, Herbert. 1960. *The new science of management decision*. New York: Harper and Row.

———. 1969. *The sciences of the artificial*. Cambridge, MA: MIT Press.

Smart, J. J. C. 1963. *Philosophy and scientific realism*. London: Routledge.

Smith, Nick. 2000. *Seeds of Opportunity: An Assessment of the Benefits, Safety and Oversight of Plant Genomics and Agricultural Biotechnology*. Report prepared for the Subcommittee on Basic Research of the House Committee on Science. 106th Cong., 2nd sess. Committee Print.

Smolin, Lee. 1997. *The life of the cosmos*. Oxford: Oxford University Press.

———. 2006. *The trouble with physics: The rise of string theory, the fall of a science, and what comes next*. New York: Houghton Mifflin.

Snyder, Laura J. 2006. *Reforming philosophy: A Victorian debate on science and society*. Chicago: University of Chicago Press.

Sober, Elliott. 1987. What is adaptationism? In *The latest on the best*, ed. J. Dupré, 105–118. Cambridge, MA: MIT Press.

———. 1997. Two outbreaks of lawlessness in recent philosophy of biology. *Philosophy of Science* 64 (4): S458–S467.

———. 1999. The multiple realizability argument against reductionism. *Philosophy of Science* 66:542–564.

Solé, Ricard, and Brian Goodwin. 2001. *Signs of life: How complexity pervades biology*. New York: Basic Books.

Solloway, M. J., and. E. J. Robertson. 1999. Early embryonic lethality in Bmp5:Bmp7 double mutant mice suggests functional redundancy within the 60A subgroup. *Development* 126:1753–1768.

Sperry, R. W. 1969. A modified concept of consciousness. *Psychological Review* 76:532–536.

———. 1991. In defense of mentalism and emergent interaction. *Journal of Mind and Behavior* 122:221–246.

Springer, M. S., J. A. W. Kirsch, and J. A. Case. 1997. The chronicle of marsupial evolution. In *Molecular evolution and adaptive radiation*, ed. T. J. Givnish and K. J. Sytsma, 129–161. New York: Cambridge University Press.

Stelling, Jörg, Uwe Sauer, Zoltan Szallasi, Francis J. Doyle III, and John Doyle. 2004. Robustness of cellular functions. *Cell* 118 (6): 675–685.

Stephan, Achim. 1997. Armchair arguments against emergentism. *Erkenntnis* 46:305–314.

Sterelny, Kim. 1996. Explanatory pluralism in evolutionary biology. *Biology and Philosophy* 11:193–214.

Streeck, Wolfgang. 2009. *Re-forming capitalism: Institutional change in the German political economy.* Oxford: Oxford University Press.

Strevens, Michael. 2003. *Bigger than chaos: Understanding complexity through probability.* Cambridge, MA: Harvard University Press.

Strohman, Richard. 2002. Maneuvering in the complex path from genotype to phenotype. *Science* 296:701–703.

Tabery, James. 2008. R. A. Fisher, Lancelot Hogben, and the origin(s) of genotype-environment interaction. *Journal of the History of Biology* 41:717–761.

Taylor, D. R., and P. K. Ingvarsson. 2003. Common features of segregation distortion in plants and animals. *Genetica* 117:27–35.

Tinbergen, Niko. 1963. On aims and methods in ethology. *Zeitschrift für Tierpsychologie* 20:410–433.

Topoff, H. 1972. Theoretical issues concerning the evolution and development of behavior in social insects. *American Zoologist* 12:385–394.

Travis, J. 1992. Scoring a technical knockout in mice. *Science* 256:1392–1394.

Tyson, John J., Katherine C. Chen, and Bela Novak. 2003. Sniffers, buzzers, toggles and blinkers: Dynamics of regulatory and signaling pathways in the cell. *Current Opinion in Cell Biology* 15:221–231.

van Swinderen, B., and J. Greenspan. 2005. Flexibility in a gene network affecting a simple behaviour in *Drosophila melanogaster*. *Genetics* 169:2151–2163.

von Wright, Georg Henrik. 1963. *The logic of preference*. Edinburgh: University of Edinburgh Press.

Vranas, Peter B. M. 2000. Gigerenzer's normative critique of Kahneman and Tversky. *Cognition* 76:179–193.

Wagner, Andreas. 1999. Causality in complex systems. *Biology and Philosophy* 14:83–101.

———. 2005. *Robustness and evolvability in living systems*. Princeton, NJ: Princeton University Press.

Waibel, Markus, Dario Floreano, Stephane Magnenat, and Laurent Keller. 2006. Division of labour and colony efficiency in social insects: Effects of interactions between genetic architecture, colony kin structure and rate of perturbations. *Proceedings of the Royal Society* B 273:1815–1823.

Walters, C. J. 1997. Challenges in adaptive management of riparian and coastal ecosystems. *Conservation Ecology*, online 1 (2): 1. http://www.consecol.org/vol1/iss2/art1.

Weinberg, Steven. 1993. *The first three minutes: A modern view of the origin of the universe.* 2nd edition. New York: Basic Books.

Weinert, R. 1995. *Laws of nature.* Berlin: Walter de Gruyter.

Weniger, G., C. Lange, and E. Irle. 2006. Abnormal size of the amygdala predicts impaired emotional memory in major depressive disorder. *Journal of Affective Disorders* 94:219–229.

West, Geoffrey, and James H. Brown. 2005. The origin of allometric scaling laws in biology from genomes to ecosystems: Towards a quantitative unifying theory of biological structure and organization. *Journal of Experimental Biology* 208:1575–1592.

Whewell, William. 1840. *The philosophy of the inductive sciences, founded upon their history.* London: John W. Parker.

Wilhere, George F. 2002. Adaptive management in habitat conservation plans. *Conservation Biology* 16 (1): 20–29.

Wilson, Edward O. 1981. *The insect societies.* Cambridge, MA: Harvard University Press.

Wimsatt, William. 1986. Forms of aggregativity. In *Human nature and natural knowledge,* ed. A. Donagan, A. N. Perovich Jr., and M. V. Wedin, 259–291. Boston: Reidel Publishing Co.

———. 1987. False models as means to truer theories. In *Neutral models in biology,* ed. M. H. Nitecki and A. Hoffman, 23–55. Oxford: Oxford University Press.

———. 2000. Emergence as non-aggregativity and the biases of reductionisms. *Foundations of Science* 5:269–297.

———. 2007. *Re-engineering philosophy of limited beings: Piecewise approximations to reality.* Cambridge, MA: Harvard University Press.

Winston, Mark L. 1987. *The biology of the honey bee.* Cambridge, MA: Harvard University Press.

Wolfer, David P. 2002. What's wrong with my mouse? Behavioral phenotyping of transgenic and knockout mice. *Genes, Brain and Behavior* 1:131.

Wolff, J. A., and J. Lederberg. 1994. An early history of gene transfer and therapy. *Human Gene Therapy* 5 (4): 469–480.

Woodward, James. 2001. Law and explanation in biology: Invariance is the kind of stability that matters. *Philosophy of Science* 68:1–20.

———. 2003. *Making things happen: A theory of causal explanation.* Oxford: Oxford University Press.

Woolgar, Steven, ed. 1988. *Knowledge and reflexivity: New frontiers in the sociology of knowledge.* Newbury Park, CA: Sage.

Worrall, J. 1989. Structural realism: The best of both worlds? *Dialectica* 43:99–124. Reprinted in *The philosophy of science,* ed. D. Papineau, 139–165. Oxford: Oxford University Press.

Wright, Sewall. 1931. Statistical methods in biology. *Journal of the American Statistical Association* 26:155–163.

Wu, F. 2006. An analysis of Bt corn's benefits and risks for national and regional policymakers considering Bt corn adoption. *International Journal of Technology and Globalisation* 2 (1/2): 115–136.

Yablo, Steven. 1999. Intrinsicness. *Philosophical Topics* 26:479–505.

Yearly, S. 1996. Nature's advocates: Putting science to work in environmental organizations. In *Misunderstanding science? The public reconstruction of science and technology*, ed. A. Irwin and B. Wynne, 172–190. Cambridge: Cambridge University Press.

Yoon, Carol Kaesuk. 2000. Altered salmon lead the way to the dinner plate, but rules lag. *New York Times*, May 1.

Zammit, S., and M. J. Owen. 2006. Stressful life events, 5-HTT genotype and risk of depression. *British Journal of Psychiatry* 188:199–201.

INDEX